RAPTORS

OF MEXICO
AND CENTRAL AMERICA

RAPTORS

OF MEXICO
AND CENTRAL AMERICA

WILLIAM S. CLARK & N. JOHN SCHMITT

WITH A FOREWORD BY LLOYD KIFF

NATIONAL AVIARY
PITTSBURGH, PA

Princeton University Press
Princeton and Oxford

Published by Princeton University Press, 41 William Street, Princeton, New Jersey 08540
In the United Kingdom: Princeton University Press, 6 Oxford Street, Woodstock, Oxfordshire OX20 1TR
press.princeton.edu

Cover illustration by N. John Schmitt

Illustration previous page: adult Roadside Hawk

ISBN 978-0-691-11649-5

Library of Congress Control Number: 2016945499

British Library Cataloging-in-Publication Data is available

This book has been composed in Chaparral Pro (text) and Arial (headings)

Printed on acid-free paper. ∞

Designed by D & N Publishing, Baydon, Wiltshire, UK

Printed in China

10 9 8 7 6 5 4 3 2 1

CONTENTS

FALCONIDS: Falconidae

LIST OF PLATES

FOREWORD

"Middle America" (Mexico and Central America) is one of the richest regions of the world for birds, and it is a virtual paradise for those particularly interested in birds of prey and their conservation. It is home to 69 species of eagles, hawks, kites, falcons, and vultures. The known distribution of neotropical raptors is surprisingly dynamic, owing mostly to large-scale habitat changes and an increasing number of capable observers. Some of the new raptors for this region are birds of open country or second-growth habitats that have somehow leapfrogged the primary rainforests of the Panamanian Darien. Others are forest birds that have expanded their ranges northward incrementally, or previously unrecorded northern species that have been detected by teams of observers at hawk watch sites. As the effects of global climate intensify, we can expect even more changes in the status of raptor species in Middle America.

Bill Clark was saved from a career in nuclear engineering and computer science by his passion for raptors. Although most regions of the world have their own respected raptor identification experts, Bill is the acknowledged global authority in this field. No other raptor has travelled so widely, or analyzed in detail the finer points of distribution, taxonomy, and identification characters of the diurnal raptor species. He has authored or co-authored 5 books on raptor identification, including his co-authored now-venerable guide to North American raptor identification in the *Peterson Field Guide* series, a similar guide to the raptors of Europe, the Middle East, and North Africa (illustrated by John Schmitt), and a field guide to raptors of the African continent that is in press. He has also published over 100 peer-reviewed publications on raptor taxonomy, identification, and hybridization, based on his field studies and examination of thousands of museum specimens. Raptors are his life. If you meet Bill, or go on one of the occasional raptor tours that he leads, be prepared. His idea of a big night out is discussing in relentless detail the subspecific status and minute plumage details of a particular population of raptors over (and after) dinner.

Among the paintings hanging on our walls is a modest black-and-white portrayal of a flock of soaring Swallow-tailed Kites. It was prepared by John Schmitt with a ballpoint pen when he was 18 years old and was quite possibly his first attempt at drawing raptors. Growing up in southern California, he had not seen the actual kites yet, but he nailed their butterfly-like mastery of the air, and the painting stands on its own merits in addition to its historical significance. Little did he (nor we) know that he would someday become one of the most respected bird artists in the world, with a special interest in raptors. Unlike many field guide artists, who are compelled to rely solely on museum specimens and photographs to create their paintings, John has actually spent a great deal of time in the field with almost all of the species portrayed here. He is a remarkably gifted observer, one who is able to grasp the gestalt of the whole bird while somehow also focusing on the patterns of individual feathers.

Interest in raptors in Middle America has intensified in recent years with the creation of specialized groups focusing on raptor migration or population monitoring in almost every country. All of the participants in these projects will want to be armed with this new field guide, which depicts and describes all of these raptors in detail and compares similar species. Further details on their behaviors, the habitats they use, and their distribution are also provided. It will be an invaluable tool for these groups as we move into the uncharted waters of global climate change and its concomitant challenges to conservationists everywhere. In particular, I hope that this excellent volume will propel certain young people into the sort of life that Bill and John have enjoyed so much.

Lloyd Kiff, July 2016
Whidbey Island
Washington

ACKNOWLEDGMENTS

This guide would not have been written were it not for our editor, Robert Kirk of Princeton University Press, who believed that we could produce it and gave us a contract. WC is most thankful to Dayton Baker, then director of the National Aviary in Pittsburgh (NAIP), who also believed in the guide project and arranged for travel grants for WC's field studies in Mexico and Central America. NAIP is a sponsor of this guide. Many people helped us with our raptor studies in the field, providing information, capturing raptors for in-hand looks, photos, and measurements, inspiration, or other support or some combination of these. We thank Bud Anderson, Roni Martinez, Rosabel Miró, Jorge Montejo, Ryan Phillips, Ernesto Ruelas, Cesar Sanchez, Sergio Seipke, Pat and Kitty Wade, Dan and Kay Wade, and Jim Zook.

We thank the curators and collection managers at the following institutions for permission to study their bird specimens: Field Museum (Chicago, Ill.); Los Angeles County Museum (Los Angeles, Calif.); Museum of Natural History, San Juan, Costa Rica; Museum of Vertebrate Zoology (Berkeley, Calif.); Natural History Museum (British Museum), Tring, England; Phelps bird collection, Caracas, Venezuela; Rancho Grande Museum, Maracay, Venezuela; University of Costa Rica bird collection; UCLA bird collection (Los Angeles, Calif.); US National Museum (Smithsonian); and Western Foundation of Vertebrate Zoology (Camarillo, Calif.).

The following photographers provided color raptor images used in the guide; the foremost was Ryan Phillips (Belize Raptor Research Institute, BRRI), and also John Afdem, Roger Ahlman, Sam Angevine, Nick Athanas, Glenn Bartley, Chris Benesh, Peter Boesman, Rick Cech, Christian Contreras Castro, Knut Eisermann, Stuart Elsom, Jörgen Fyhr, Jim Gallagher (Sea & Sage Audubon Society), Michel Giraud-Audine, Mark Harper, Ned Harris, Antonio Hidalgo, Chris Jiménez, Richard Kuenn (BRRI), Greg Lasley, Jerry Liguori, Jim Lish, Lior Kislev, Sean McCann, Richard Pavek, Jimmy Paz, Patty Rainey, Jean-Louis Rousselle, Cesar Sanchez, Sergio Seipke, Yeray Seminario, Brian Sullivan, Kevan Sunderland, Clay Taylor, Brian Wheeler, Andrew Whittacre, Chris Wood, Barry Zimmer, Kevin Zimmer, and Jim Zipp.

Oliver Komar, Ryan Phillips, John van Dort, and Jim Zook provided helpful comments on the draft range maps. And we would like to express our gratitude to Diana Abdella for her generous dedication of time and skills in producing the final range maps.

INTRODUCTION

Mexico and Central America have a wide variety of diurnal raptors, due to their connection to both North America and South America and a broad diversity of habitats from temperate to tropical. Many of these species are migrants from North America that pass through on migration or winter in this area. Sixty-nine species plus one subspecies are covered herein.

The aim of this raptor field guide is to present the latest information on tried and proven field marks for the field identification (ID) of diurnal raptors in Mexico and Central America. William S. Clark (WC) has previously written a field guide to the raptors of North America (Clark and Wheeler 2001). WC decided to write this guide after several trips to this area, when he noted that most of the bird field guides he used there did not cover the diurnal raptors well. Especially, they failed to show accurately the shapes of raptors' wings and tails and lacked field marks to distinguish them. Also, he already knew many of these raptors, as many of them were covered in the North American guide and from his travels in Latin America. WC authored a similar raptor field guide for Europe (Clark 1999), which was illustrated by N. John Schmitt (JS).

Our research as we set out to produce this guide was in five main areas. First, we took many photographs of raptors in the field, both perched and flying, as well as in zoos and rehabilitation facilities, and of raptors in hand captured for banding in many locations. WC studied many videos of raptors as well. Next, WC studied specimens of raptors in numerous museums. Third, WC gathered and studied the literature on neotropical raptor ID and related subjects. Further, WC studied carefully the plumages of many raptors that had been captured for banding or in rehabilitation and took measurements of them to get more accurate data on wingspan, length, and weight. Finally, WC spent much time in the field in many locations in Mexico and Central America, field-testing the field marks presented in this guide. Many of these field marks have been published by him and others, but there is much new unpublished information reported herein for the first time.

JS has illustrated birds, especially raptors, in many field guides, including North America, India, and Peru. Further, he has a great deal of experience with raptors through field and museum work.

We also have spent considerable time together observing raptors in museums and in the field.

The field marks presented herein will enable a person who is viewing a raptor anywhere in Mexico and Central America in good light with good to excellent optical equipment to identify all but the most unusual individuals.

COLOR PLATES

All of the 32 color plates were painted by JS expressly for this guide. We aimed for one plate per species where necessary, but in order to illustrate all species on 32 plates, up to three species are illustrated on certain plates. In all cases, the postures and positions of the birds are chosen to facilitate easy comparison across species and to help the eye navigate across possible identification contenders; for example, adult females or juvenile males are shown in the same posture and position on the layout for each species for direct comparison. Flying figures are shown at least once in a full-soar posture, with the exception of species that do not show this behavior. Other figures showing typical gliding and faster-flying profiles are included where possible. Within large, similar-sized groupings—for example, falcons and hawks—we closely adhered to a consistent scale based on wingspan and body-length measurements. The plates are ordered by species in more or less taxonomic order; however, some species are placed out of order for space or comparison reasons, for example. The last plate consists of three vagrant raptors that barely reach eastern Panama (two) or northern Mexico (one) and one introduced in northwestern Mexico.

The plates were prepared as a joint effort between us. WC provided JS with a list of raptors on each plate by sex, age, color morph, and attitude (perched and shown from front or back or side, soaring, flying below, above, or head-on) for the species to be illustrated and short instructions on which plumages to show. WC also provided JS additional reference material, such as color photographs of raptors flying, perched, in hand, and in cages; and articles on raptors to supplement RD's reference material. JS then scanned a pen-and-ink outline of the plate, showing each figure in its exact size, shape, and position; he occasionally suggested additional figures. WC reviewed each outline for accuracy of wing and tail shapes and proportions among species and sex classes; often a few minor adjustments were made on the initial outline. After agreement on the outline, JS painted in the color figures and sent WC a color copy of the plate. JS added representative habitat backgrounds in pencil or color wash. WC then made comments on the colored figures for minor alterations on some plates.

PLATE CAPTIONS

Each plate has on the opposing page a list of figures by species that describe field marks shown on each figure. Treatment of each species begins with a title (the common and scientific names), a pointer to the species account in the text, and then a general summary of identification points and the list of figures. Species are numbered from top to bottom on the plates, and figures for each species are lettered, beginning with *a*, from left to right and top to bottom. Each is labeled as to age and sex; in some instances, color morphs or subspecies are labeled. Then follow the traits shown in this figure, sometimes with pointers to similar figures of other species and how they differ.

TAXONOMY

The taxonomy and order used was that of the American Ornithologists' Union (AOU) and South American Ornithologists' Union (SAOU), except that many taxonomic changes are made based on recently published papers. We have gone to considerable lengths to cover all the significant geographic variation in the looks of these readily identifiable taxa that are known at this point in time.

COMMON NAMES

The common names used are those of the AOU, with some exceptions. Spanish common names follow the recommendations of Seipke et al. (2007).

SPECIES ACCOUNTS

The main text of this guide consists of the species accounts, one for each species of diurnal raptor that has been recorded in Mexico and Central America. Species accounts are written in the following format.

SPECIES HEADING The heading consists of the English common and scientific names, with pointers listing the color plate number(s) for the species. The Spanish name appears on the following line.

IDENTIFICATION SUMMARY This is a list of field marks or general characters that apply to all individuals of the species, as well differences in sexes and ages. Next are detailed descriptions of each regularly occurring plumage variation by age, sex, subspecies, and color morph, including eye, cere, and leg or toe colors. These descriptions are for raptors viewed at close hand.

MEASUREMENTS Measurements of wingspan, total length, and weight are given. Data on the first two in many references are incorrect; thus WC undertook to take these measurements from live raptors whenever possible. This was supplemented by data taken from museum specimen tags, as they are for the most part almost the same as data taken by WC on live birds. Total length data given are taken from the top of the head to the tip of the tail, as this is how the raptors are seen when perched. These measurements are a bit smaller than those taken on specimens, which are from the tip of the beak with the head pointed upward to the tip of the tail. Weight information in the literature is usually accurate; nevertheless, we used mostly data taken by us and our colleagues, supplemented by museum tag data, especially for those species that we did not handle as live birds. Ferguson-Lees and Christie (2001) was very helpful.

TAXONOMY AND GEOGRAPHIC VARIATION The subspecies that occur in Mexico and Central America of each polytypic species are given, along with range of each. This section includes comments on recent taxonomic changes, with references.

SIMILAR SPECIES Species that can be confused in the field with the species in each account are listed, along with the plate number or numbers on which each is illustrated, followed by the field marks that separate them. These are given only once; thus, for some accounts, after a short sentence on similarities between two species, readers will find the note "See under that species for distinctions."

STATUS AND DISTRIBUTION The breeding range or the permanent range and the status within that range are presented, along with information on migration and dispersal. Any population declines are noted, with reasons for these given, if known.

HABITAT All of the habitats in which a species regularly occurs.

BEHAVIOR Any behavior that will aid in field identification is described. Also described are hunting methods and main prey. A brief discussion of nesting substrate and display flights follows. Three methods of flight are described: how the wing beats appear in powered flight and how the wings are held in soaring and gliding flight. If the species hovers, this is also noted. On some species—for example, the kites—particular flight methods may be distinctive and are described.

MOLT This section is a description of when and how raptors molt their feathers. Some adults complete their molt annually on the breeding grounds before and after breeding (molt is usually suspended while breeding). Others, especially trans-equatorial migrants, molt some feathers on the breeding grounds but suspend molt during the autumn migration and complete their molt on the winter grounds. Some of the larger species do not replace all of their feathers in one year; usually two years or more are required to renew all feathers. Also discussed is the molt of juveniles, as it differs from that of adults in some species. Some begin molt in their first spring, when they are almost a year old, whereas others, particularly trans-equatorial migrants, begin molt earlier on the winter grounds and replace some or most of the body feathers and sometimes also a few tail and flight feathers. Molt is usually suspended during migration.

Age terminology is related to molts. All species are in juvenile plumage when they leave the nest. Many species undergo one annual molt into adult plumage; others, particularly the larger eagles and buzzards, require three or four annual molts to reach adult plumage, with several immature plumages in between. See the Raptor Glossary, below, for the definitions of the immature plumages: second through fourth plumages (Basic II–Basic IV). See Clark and Pyle (2015) for a more detailed description of age and molt terminology of raptors.

Understanding the molt sequences of the flight feathers is important to properly age large raptors in immature plumages. See Edelstam (1984), Miller (1941), and Jollie (1947) for a thorough discussion of molt in raptors. In accipitrid raptors, primary molt begins with number 1 (P1), the innermost feather, and proceeds outward in sequence. In falconids, primary molt begins with number 4 (P4) and proceeds both inward and outward. When all primaries are not replaced in one year, then the sequence continues where it left off when molt is restarted the next spring. However, in accipitrids, molt of primaries occurs again during this molt season, beginning a new wave molt. (See Clark [2004a] for a full discussion of using wave molt in ageing raptors.) The molt of secondaries in accipitrid juveniles begins at three molt centers, S1, S5, and S13–16, depending on size. It proceeds inward from S1 and S5 and outward from S13 to S16 (also inward to replace the tertiaries). In species that do not replace all secondaries every year, the pattern of molt is apparently random. In falconids, secondary molt begins with S4 or S5 and proceeds both outward and inward. Other publications that discuss molt are cited in some species accounts.

Molt of the tail feathers is sometimes irregular; however, it usually begins with T1, the central pair. The usual sequence after that is T2, T3, T6, T4, and then T5. However, there is a lot of variation in this order and even some asymmetrical molt, particularly beginning with the second annual molt.

Many species undergo a pre-formative molt of some to many feathers soon after fledging and before the onset or the second pre-basic molt, usually of feathers on the upper back, upper wing coverts, upper breast, and neck. In some species—for example, White-tailed Kite and American Kestrel—this molt is extensive. Juveniles of Mississippi Kites undergo a more or less complete body molt while in South America, whereas juveniles of Swainson's Hawk do not. See Pyle (2005) for a discussion of pre-formative molt in American raptors.

DESCRIPTION After a summary paragraph, there follows a detailed discussion of each age class, including sex differences and color morphs.

FINE POINTS More detailed information on field identification is presented.

UNUSUAL PLUMAGES Any of a variety of abnormal plumages that have occurred in the species are included in this section. Abnormal plumage types are: albinism, partial albinism, amelanism, partial amelanism, melanism, and partial melanism. Albinism is the complete absence of any pigmentation in the plumage, soft parts, and eyes and is extremely rare in raptors. Partial albinism is when several to most or even all feathers lack pigmentation and are white, but the eyes and some soft parts are a normal color. Amelanism is a condition in which the amount of melanin in many or all feathers is reduced such that normally brown feathers appear as a café-au-lait hue (also called *dilute plumage*, *leucism*, and *schizochroism*). With partial amelanism only some feathers are affected or the amount of pigment reduction is not uniform. Melanism results from an excess of dark pigment such that the raptor is completely dark brown on the body and coverts (but, interestingly enough, not always on the flight and tail feathers). But note that the term *melanism* applies only to abnormal plumages; it is called a "dark color morph" when it occurs regularly. In partial melanism only parts of the raptor have excess pigmentation.

HYBRIDS Any known hybrids are listed.

ETYMOLOGY The origins of the common and scientific names are given except for the obvious ones. In some cases, alternate explanations are given.

REFERENCES The citations in the section are summarized in alphabetical order. A list of references is included as an appendix.

RANGE MAPS

Range maps are provided to give the reader a general idea of the distribution of each species. Finer details are available in the bird field guides for each country. Range maps are not provided for vagrant and introduced raptors. The colors on the maps are shown in the sample below and indicate a species' presence as follows:

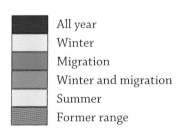

All year
Winter
Migration
Winter and migration
Summer
Former range

ISLAND SPECIES AND SUBSPECIES

We did not include species and subspecies on islands off the mainland, as that would have increased the number covered with little benefit to the raptor watcher on the mainland.

GLOSSARY

A glossary is included after the introduction that explains the terms used in the guide.

HOW TO USE THIS GUIDE

As a field guide, this book is to be carried in the field during bird-watching, especially raptor-watching. However, it is also to be used as a reference book, to be read and studied at home in preparation for field work.

When a raptor is seen well and its field marks noted, even a beginner can use this guide to identify it correctly. Taking photographs for later study using this guide will also lead to correct IDs.

Please consider submitting your observations of raptors in Mexico and Central America to eBird (www.ebird.org/content/ebird).

HELPFUL FACTS FOR RAPTOR FIELD IDENTIFICATION

1. Juveniles in fresh plumage usually show clean, unflawed uniform plumage and pale tips on flight and tail feathers and greater wing coverts; the latter form narrow pale lines on the wings. These tips usually wear off in winter.

2. Non-juvenile raptors in summer and early autumn often show signs of molt (gaps, notches, etc.) in wings and tail, including uneven trailing edges of wings and tip of tail, and show a mix of new fresh and old faded feathers.

3. Many raptors have different lengths of secondaries in juvenile and older plumages. Some have longer secondaries as juveniles (the wings appear wider); others have shorter ones (the wings appear narrower). Most raptors have a longer tail in juvenile plumage.

4. The flight feathers of juvenile raptors, especially secondaries, often show very pointed tips, whereas adult flight feathers have square or rounded tips. This results in a noticeably serrated trailing edge to the wings and sometimes the tail of juvenile birds.

5. Flight and tail feathers are darker on the uppersides as compared to the undersides, and darker on the outer webs compared to the inner webs. Hence, flight and tail feathers appear darker on the uppersides.

6. Some raptors show pale areas on back-lighted underwings (windows or panels).

7. Rufous underparts in fresh plumage of juvenile raptors—for example, Red-tailed Hawks and Ferruginous Hawks—usually fade due to sunshine and weather to buffy, creamy, or even whitish a few months after fledging.

8. Raptors that show rounded wingtips when soaring often show somewhat pointed wingtips when gliding.

9. There are two somewhat separate problems in raptor identification: perched and flying. The field marks used for each may be different. For example, wing shapes and underwing patterns of soaring raptors are not visible on the same raptors perched, when the relative position of wingtip and tail tip can be field marks.

10. When raptors are seen at close range, many details and shadings of color are noticeable; the same raptor seen at a distance appears to show only light and dark areas, with loss of definition and color.

11. Raptors, and other birds as well, often appear to have different colors under differing lighting conditions. All flying raptors appear darker against whitish skies, for example.

RAPTOR GLOSSARY

Adult plumage. The final breeding plumage of a raptor. Also called *Definitive Basic plumage*.

Albinism. Rare abnormal condition in which all of the feathers, beak, cere, talons, and eyes of a bird are without pigmentation. Feathers are white, and beak, cere, and talons are ivory. See Partial albinism, Partial amelanism, and Amelanism.

Allopatric. Said of two taxa whose breeding ranges do not overlap.

Allopreen. The act of preening or grooming the skin or feathers of another bird, often a mate.

Amelanism. An abnormal plumage in which the dark colors are replaced by a lighter, usually creamy color (but not white); caused by a reduction in pigmentation. See Partial amelanism, Albinism, and Partial albinism.

Arched wings. Wings held with wrists above the body and with wingtips lower.

Arm. The inner half of the wing of a flying bird.

Auriculars. The feathers covering the ear. See fig. 3c.

Axillaries or **axillars.** Feathers at the base of the underwing, also called the *armpit* or *wing pit*. See fig. 1.

Back. See fig. 2.

Band. A stripe of contrasting color, usually in the remiges and tail. See Tail banding.

Bare parts. Unfeathered parts of a bird, usually the beak, cere, lores, eye-ring, and, for many species, legs.

Barring. Series of narrow bands anywhere on the body or coverts that contrast with the background color. See fig. 1.

Basic plumages. Basic II is an immature raptor's second plumage, Basic III the third, and so on. Definitive Basic plumage is the final breeding plumage of a raptor.

Beak. Same as Bill. See fig 3c.

Belly. See fig. 3b.

Bib. Pattern of uniformly dark breast that contrasts with paler belly. See fig. 1.

Bill. Bony projection that is part of a raptor's jaw. See also Mandibles. See fig 3c.

Breast. See fig 3b.

Buteonine. Any accipitrid raptor in the genus *Buteo* and the closely related genera: *Buteogallus*, *Geranoaetus*, *Leucopternis*, *Rupornis*, *Morphnarchus*, *Parabuteo*, *Pseudastur*, and *Cryptoleucopteryx*.

Buzzard. A misnomer applied to Turkey and Black Vultures in North America. Common name of buteonines outside of North America.

Caching. The act of a raptor storing prey, usually excess, for later retrieval, often in holes in trees, poles, or cliff faces.

Carpal. The underwing at and just beyond the wrist, usually composed of all primary underwing coverts. See fig. 1.

Cere. A small area of often colorful bare skin enclosing the nostril (nare) at the base of the upper mandible. See fig. 3c.

Cheek. See fig. 3c.

Chin. Area directly under the beak. See fig. 3c.

Collar. A pale band across the hind neck.

Coverts. The small feathers covering the bases of the flight feathers and tail both above and below. See figs. 1 and 2. See also Wing coverts and Tail coverts.

Crest. Elongated feathers on the rear crown, usually erectable. See fig. 1.

Figure 1

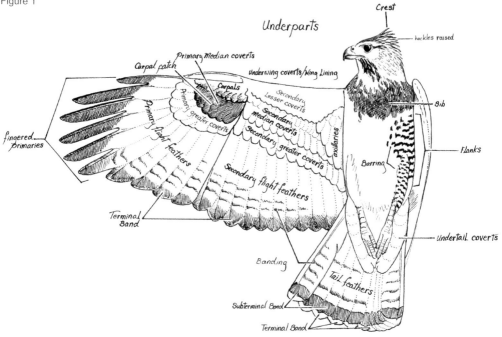

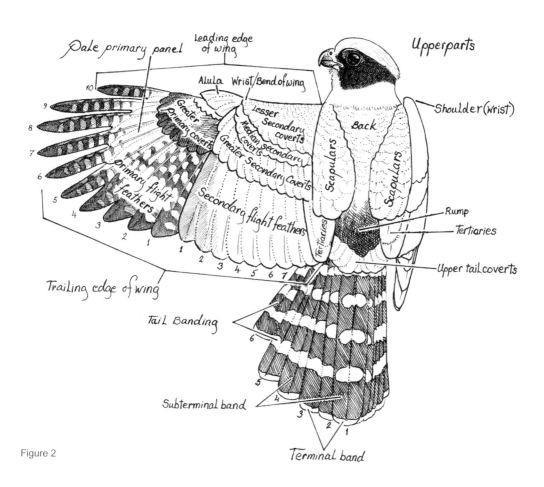

Figure 2

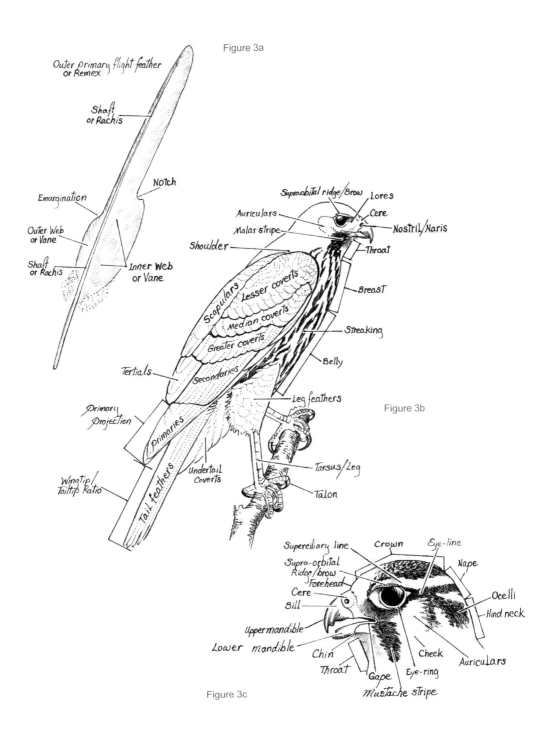

Figure 3a

Outer primary flight feather or Remex

Shaft or Rachis

Notch

Emargination

Outer Web or Vane

Shaft or Rachis

Inner Web or Vane

Supraobital ridge/Brow Lores

Auriculars Cere

Malar stripe Nostril/Naris

Shoulder Throat

Scapulars Lesser coverts Breast

Median coverts

Greater coverts Streaking

Belly

Tertials Secondaries

Primary Projection

Primaries

Leg feathers Figure 3b

Wingtip/Tailtip Ratio

Tail feathers

Undertail Coverts

Tarsus/Leg

Talon

Superciliary line Crown Eye-line

Supra-orbital Ridge/brow Nape

Forehead

Cere Ocelli

Bill Hind neck

Upper mandible

Lower mandible Chin Cheek

Throat Auriculars

Gape Eye-ring

Mustache stripe

Figure 3c

Crop. A wide area of the esophagus in the neck of most raptors used to store food for a short while for predigestion. When full of food, a crop often shows a variably colored bare skin area on the front of the lower neck in some species, especially vultures and eagles.

Crown. Top of head. See fig. 3c.

Dark morph. Color morph of a bird in which the underparts and underwing and tail coverts are overall dark, usually dark brown. See also Melanism.

Dihedral. The shape when a bird holds its wings above the horizontal (positive dihedral), further defined as:

(1) Strong dihedral: wings held more than 15 degrees above level;

19

(2) Medium dihedral: wings held between 5 and 15 degrees above level;

(3) Shallow dihedral: wings held between 0 and 5 degrees above level;

(4) Modified dihedral: wings held between 5 and 15 degrees above level, but held nearly level from wrist to tip.

Negative dihedral is when wings are depressed below the horizontal, as in Double-toothed Kites and some falcons.

Dilute plumage. Another term for Amelanism and Partial amelanism.

Dimorphic. Relating to a species that has two color morphs or two sizes (usually between sexes).

Disjunct. Said of two taxa whose distributions do not overlap.

Dispersal. The movement of recently fledged juveniles away from their nesting area.

Diurnal. Active in the daytime.

Dorsal. The uppersides of a bird.

Emargination. An abrupt narrowing on the outer web of an outer primary; see also Notch. See fig. 3a.

Extirpated. Said of a bird that no longer breeds or occurs in an area.

Eye-line. Dark line through the eye, often visible only behind the eye. Also called *eye-stripe*. See fig. 3c.

Eye-ring. The bare skin around the eye, somewhat wider in front of the eye in falcons. See fig. 3c. Also called *orbital ring*.

Face skin. The lores when bare of feathers.

Facial disk. A saucer-shaped disk of feathers around the face, thought to direct sound to the ears. Most visible on harriers (also on owls).

Feather edge. The edges of a feather. Pale edges can give the effect of streaking or barring.

Feather fringe. The complete circumference of a feather. Pale fringes usually give a scalloped appearance.

Field marks. The characters of a bird that serve to distinguish it from similar species.

Fingers. The tips of outer primaries that have notches and emarginations.

Flanks. The sides of the under body. See fig. 1.

Fledgling. A raptor that has just left the nest, fledged.

Flight feathers. The primaries and secondaries (remiges) and tail feathers (rectrices).

Forehead. See fig. 3c.

Forked tail. Tail in which the outer tail feathers are longer than the central ones, resulting in a variably sized notch on the tip. Also called *swallow tail* or *kite tail*.

Fourth plumage. Fourth non-adult plumage of raptors that take at least four years to reach adult plumage, especially large eagles. Also called Basic IV.

Fifth plumage. First adult plumage of raptors that take at least four years to reach adult plumage, especially large eagles, when some immature characters are still noticeable. Also called Basic V.

Gape. The soft skin, often colorful, at the base of the mandibles (bill); corners of the mouth. See fig. 3c.

Glide. Flight attitude of a bird when it is coasting downward. The wingtips are pulled back, more so for steeper angles of descent, usually with the tail closed.

Gregarious. Term for birds that gather in groups or flocks, especially for feeding, roosting, or migration. Such birds are also called *social*.

Gular stripe. Vertical stripe on the chin and throat.

Hackles. Erectable feathers on the nape.

Hand. The outer half of the extended wing on a flying bird.

Hawk. A raptor of the genus *Accipiter* and closely related genera, but also used generically for all diurnal raptors.

Hind neck. See fig. 3c.

Hover. To remain in a fixed place, facing into the wind, by flapping. More properly called *wind hover*.

Hybrid. A bird with parents of different species.

Immature. Non-specific term that means "not adult." It is used in some field guides to mean "juvenile" and in others for the stages between juvenile and adult. Used herein to mean not an adult when the actual age is not apparent.

Inner web of feather. See fig. 3a.

Juvenile plumage. First complete plumage, acquired in the nest. Usually differs from the plumage of adults.

Kite. A verb meaning to remain in a fixed place in moving air on motionless wings.

Leading edge of wing. See fig. 2.

Leg. See fig. 3b.

Leg feathers. Feathers of the upper legs. Also called *leggings* and *trousers*. See fig. 3b.

Length. For a perched bird, the distance from the top of the head to the tip of the tail.

Leucistic. A vague term used often for *amelanism* and *partial amelanism* but also *partial albinism* and *albinism*. It literally means "whitish."

Light morph. Color morph of a bird in which the underparts and underwing and tail coverts are white or whitish.

Lores. The area of the sides of the face between the eye and the beak. See fig. 3b.

Lower mandible. See fig. 3c.

Malar stripe. A dark mark on the cheek that begins at the base of the lower mandible. Many buteonines show a malar stripe. See fig. 3b.

Mandibles. The upper and lower halves of the bill.

Marginal wing coverts. Tiny feathers at the bend of the wing covering the bony leading edge of primaries.

Melanism. Rare abnormal condition in which a raptor's feathers are darker due to excess dark pigmentation. See also Dark morph.

Migratory. Said of a bird that spends parts of the year in different locations.

Molt. Means by which a bird replaces its feathers.

Monotypic. Having only one subspecies.

Morph. Term used for recognizably different forms of a species, usually color-related. Color morphs are dark, rufous, and light. See Phase.

Mustache stripe. A dark mark that begins under the eye. Most falcons show a mustache stripe. See fig. 3c.

Nape. The back of the head. See fig. 3c.

Nocturnal. Active at night.

Nostril. Opening on the cere for breathing. Also called *nare* or *naris*. See fig. 3b.

Notch. Abrupt narrowing of the inner web of an outer primary. See also Emargination. See fig. 3a. *Notch* is also used to describe the upper mandibles of falcons.

Ocelli. Dark or light spots on the nape and hind neck that resemble eyes. See fig. 3c.

Older immature. One or more age classes between juvenile and adult.

Outer web of a feather. See fig. 3a.

Parapatric. Said of two taxa whose breeding ranges do not overlap but are adjacent.

Partial albinism. Abnormal condition in which a few to almost all feathers of a raptor lack pigmentation and appear white. Feathers can be normal, entirely white, or partly white and partly normal. Eyes, cere, and beak colors are usually normal; some talons may be ivory. See Albinism and Dilute plumage.

Partial amelanism. An abnormal plumage in which the dark colors of some but not all feathers are replaced by a lighter, usually creamy color (but not white); caused by a reduction in pigmentation.

Patagium (adjective: patagial). The triangular area between the wrist and the body on the underwing of a bird.

Phase. Term formerly used for *color morph. Phase* implies a temporary condition; color morphs are permanent. See Morph.

Plumage. All of the feathers of a bird collectively.

Polymorphism. Relating to species that have more than two color morphs.

Pre-formative molt. A partial molt of juvenile raptors in which a few to many feathers on the back, breast, and upperwing coverts are replaced not long after fledging.

Primaries. The outer remiges (hands). See fig. 1.

Primary panel. A light area in the primaries, usually more visible from below when the wing is backlighted. Also called *window*. See fig. 2.

Primary projection. The distance on the folded wing of perched birds from the tip of the longest primary to the tips of the folded secondaries. See fig. 3b.

Raptor. Any bird of prey; any member of the Falconiformes or Strigiformes, although sometimes used to refer only to the diurnal birds of prey and not owls.

Rectrices (singular: rectrix). Tail feathers. See figs. 1 and 2.

Remiges (singular: remex). Primaries and secondaries. See fig. 1.

Ruff. Elongated feathers circling the sides and rear of the lower neck, shown by the larger vultures.

Rufous morph. Color morph in which the underparts, underwing, and undertail feathers are rufous, sometimes only in adult plumage.

Rump. The lowest area of the back. See fig. 2.

Scapulars. Row of feathers between the back and the upperwing coverts of each wing. See fig. 2.

Scavenger. Raptor that eats carrion, garbage, offal, and other animal waste.

Second plumage. Immature plumage that follows the juvenile plumage on raptors that require more than one annual molt to reach adult plumage. Also called Basic II.

Secondaries. The inner remiges from the wrist to the body. See fig. 1.

Sedentary. Said of birds that remain all year in the same location. Not migratory.

Shaft of a feather. Also called *rachis*. See fig. 3a.

Shoulder. The wrist and bend (top) of the folded wing on a perched raptor. See fig. 3b.

Soar. The flight attitude of a bird with the wings, and usually the tail, fully spread. Used to gain altitude in rising air columns.

Social. See Gregarious.

Spotting. Round, dark markings usually, but not always, on the underparts.

Streaking. Vertical dark markings, usually on the underparts. See fig. 3b.

Subadult. Vague age class that has been defined in at least three ways, and thus should be avoided, as clear terms exist for all age classes.

Subterminal band. Last dark band nearest the pale tips of the tail feathers. See fig. 1.

Superciliary line. Contrasting (usually pale) line above the eye. Also called *supercilium* (singular) and *supercilia* (plural). See fig. 3c.

Supra-orbital ridge. Bony brow projection over the eye, giving raptors their fierce appearance. See fig. 3c.

Sympatric. Said of two taxa whose breeding ranges overlap.

Tail. See figs. 1 and 2.

Tail banding. See fig. 2.

Tail coverts. Feathers that cover the bases of the tail, above and below. See also Uppertail coverts and Undertail coverts. See figs. 1 and 2.

Tail feathers. Same as Rectrices. See figs. 1 and 2.

Talon. The claws of a raptor. See fig. 3b.

Talon-grappling. Interactive behavior between two flying raptors that lock feet and tumble with their wings extended.

Tarsus. Leg above the toes to the next joint; may or may not be covered by feathers. See fig. 3b.

Terminal band. Last dark band on the tips of the flight feathers. See fig. 2.

Tertiaries. Innermost short secondaries; also called *tertials*. See fig. 2.

Third plumage. Plumage that follows the second plumage (Basic II) on large raptors that take more than two years to reach adult plumage. Also called Basic III.

Throat. See fig. 3c.

Trailing edge of wing. See fig. 2.

Trousers. Long feathers on the upper legs. Also called *leg feathers* and *leggings*. See fig. 3b.

Underparts. Breast, belly, and undertail coverts. See figs. 1 and 2.

Undertail coverts. See fig. 1.

Underwing. The underside of the open wing.

Underwing coverts. Also called *wing linings*. See fig. 1.

Undulating flight. Bird flight that is up and down like a sine wave.

Upper mandible. See fig. 3c.

Uppertail coverts. See fig. 2.

Upperwing coverts. Primary and secondary coverts. See fig. 2.

Upperparts. Back and upper wing and upper tail coverts. See fig. 2.

Vagrant. A bird that occurs outside of its normal range.

Ventral side. The underside of a bird.

Wave molt. Molt of the primaries occurring in more than one location; shown on raptors that do not molt all 10 primaries annually.

Weight. The body mass of a bird.

Wing coverts. Feathers that cover the bases of the flight feathers. The three sets are: lesser, median, and greater. Further divided into coverts for primaries and secondaries. See also Upperwing coverts and Underwing coverts.

Wing linings. Underwing coverts. See fig. 1.

Wing loading. Weight divided by the wing area; a measure of the buoyancy of flight. The lower the wing loading, the more buoyant the flight.

Wingspan. Distance between the wingtips with the wings fully extended.

Wingtip extension. The distance the wingtips extend beyond the tail tip on perched raptors.

Wrist. The bend of the wing. See fig. 2.

Wrist comma. Comma-shaped mark, usually dark; the dark tips of the primary underwing coverts; seen on the underwings of many buteos.

PLATE 1

TURKEY VULTURE AND LESSER YELLOW-HEADED VULTURE

Two similar large aerial scavengers. Both locate carrion by smell and sight while in flight.

1 TURKEY VULTURE

Cathartes aura PAGE 91

Large brownish-black vulture that occurs commonly and is widespread in most habitats. Adults and juveniles similar. Wingtips reach the tail tips on perched vultures. Legs are pale gray to pinkish.

a. Adult. Head is red, beak is pale yellow, and upperparts are brownish-black.

b. Juvenile. Similar to adult, but head and beak are dark; the beak has a pale base that expands with age. Brownish feather edges on upperparts are wider than adults'.

c. Adult and juvenile. Resident vultures in Costa Rica and Panama have whitish napes. They are a bit smaller than northern vultures that winter there.

d. Juvenile. Like adult but with dark head and beak.

e. Adult. Head is completely red. Underwings are two-toned: silvery remiges contrast with dark underwing coverts.

f. Adult. Upperparts are brownish-black. Note red head. Pale primary patches are less distinctive than those of 2, lacking bright, distinct, pale shaft streaks. Note that the inner three primaries are dark (newer).

g. Head-on. Soars and glides with wings in a dihedral.

2 LESSER YELLOW-HEADED VULTURE

Cathartes burrovianus PAGE 94

Large black vulture that occurs locally in wet open areas. Adults and juveniles similar. Wingtips extend far beyond tail tip on perched vultures. Legs are pale gray to pinkish.

a. Adult. Head often appears completely yellow.

b. Adult. Head is yellow, red, and pale blue and can appear mostly red when vultures are excited. Beak is pale yellow. Upperparts are black. Note pale shaft streaks on outer primaries.

c. Juvenile. Similar to adult, but head is pale gray, nape is covered with whitish down, beak is dark, and upperparts are brownish-black. Note bright pale shaft streaks on outer primaries.

d. Adult. Similar to Turkey Vulture, but wings are proportionally longer and narrower. Note the yellow head.

e. Adult. Upperparts are black, and head appears yellow. Pale primary panels show bright pale shaft streaks. Note the yellow head with bluish crown and newer dark inner three primaries.

f. Juvenile. Similar to adults, but with dark beak and pale gray head.

g. Head-on. Soars and glides with wings in a dihedral.

PLATE 2

KING VULTURE AND BLACK VULTURE

Two scavengers, one very large and one large. Both locate carrion by sight.

1 KING VULTURE

Sarcoramphus papa PAGE 99

Very large vulture of forests. Adults and older immatures have a distinct pied plumage and colorful heads and necks. Younger immatures are blackish. Their relatively longer wings aren't obviously so because they are broad, especially near the body, which results in the tail tip appearing short. Wingtips extend a bit beyond the tail tip on perched vultures. Legs are pale gray to whitish.

a. Adult. White body and wing linings and black flight and tail feathers are distinctive. Tapered wings appear somewhat triangular. Colorful head and neck are usually not seen on flying vultures, unless seen near.

b. Oldest immature. Birds that are almost adults show some black spotting in the wing linings.

c. Juvenile. Overall dark dull black. Wing linings are variably mottled white and blackish, axillaries white smudged with blackish. Juveniles have pointed tips on secondaries. Dusky-pinkish crop is noticeable when full. Note black head.

d. Adult. White body and wing linings and black flight and tail feathers are distinctive. Tapered wings appear somewhat triangular. Colorful head and neck are usually not seen on flying vultures, unless seen near. Mantle usually shows a peachy bloom.

e. Juvenile. Uppersides appear uniformly dark. Juveniles have pointed tips on secondaries. They appear short-tailed because of broad wings.

f. Head-on. They soar with wings level or slightly raised, but glide with wings slightly cupped.

g. Second plumage. Much like juvenile, but dark wattle is longer, head and neck are more colorful, secondaries show a frosty cast, and belly, leg feathers, and undertail coverts often show more white.

h. Older immature. Head and neck are becoming adult-like, with small corrugations. Yellow wattle is short and erect. Underparts are now white, but upperwing coverts and ruff are dark.

i. Oldest immature. Birds that are almost adults show variable black spotting on the back and upperwing coverts. Yellow wattle is small and erect.

j. Adult. Colorful head and black and white plumage are unmistakable. Long yellow wattle hangs loosely over the side of the bill. Note the gray ruff.

k. Juvenile. Overall dark sooty-black, with blackish head, beak, and neck. Smooth dark head is covered with fine blackish down, and short, black wattle over the cere is erect. Belly and leg feathers and undertail coverts occasionally show some whitish smudges.

2 BLACK VULTURE

Coragyps atratus PAGE 88

Black, bare-headed vulture with short, square-cornered tail. Flies with quick shallow wing beats of stiff wings. Legs are pale gray to whitish.

a. Head-on. They soar with wings slightly raised.

b. Adult. Overall black, with pale gray primary panels and legs; legs extend to tail tip.

c. Adult. Overall black, with pale gray primary panels. Noticeable head projection and short, square-cornered tail.

d. Younger adult. Like older adult, but with less wrinkled head.

e. Juvenile. Juveniles have smooth blackish head covered with fine black fuzzy down. Feathers extend farther up nape than on adults.

f. Adult. Overall black, except for gray wrinkled featherless head, yellowish tip of beak, and pale gray legs; wingtips extend to tail tip.

PLATE 3

BALD EAGLE AND OSPREY

Two raptors usually associated with water, one large and the other even larger.

▋ BALD EAGLE

Haliaeetus leucocephalus PAGE 135

Very large, long-winged raptor. Immatures show white axillars. Wings are proportionally long. Wingtips reach or almost reach the tail tip on perched eagles.

a. Adult. White head, neck, tail, and tail coverts are diagnostic. Trailing edge of wings appears parallel to leading edge, giving a board-like look.

b. Head-on. Glides with wings level or with tips drooped a bit.

c. Basic III. Basic II and III show dark breast and white belly and axillars and dark mottling on white wing linings. Basic III has orangish beak, whitish throat and cheeks, and smooth trailing edge of wings. Some can show one or two longer pointed retained juvenile secondaries.

d. Basic II. Basic II and III show dark breast and white belly and axillars. Basic II has dark beak, dark throat and cheeks, and ragged trailing edge of wings, a mix of longer pointed juvenile and shorter, blunt-tipped replacement secondaries.

e. Juvenile. Dark underparts with tawny belly, all dark head, and serrated tips of remiges are distinctive. All immatures show white axillars and dark edges to undertails.

f. Basic II. Basic II and III show white triangle on the back. Basic II has ragged trailing edges of the wings; smooth on Basic III. Dark wing coverts show white mottling.

g. Juvenile. Two-toned upperwings—tawny coverts contrasting with dark brown remiges—and lack of white triangle distinguish this plumage. Uppertail is often completely dark.

h. Basic V. Almost adult, but with some black spots and narrow streaks on head and black tips on some tail feathers and undertail coverts.

i. Adult. White head, neck, tail, and tail coverts are diagnostic. Eyes are whitish, and cere and beak are bright orange. Dark brown body feathers show paler feather edges.

j. Basic III. Like Basic II with dark breast and white belly, but head shows white superciliary lines, white cheeks, and resulting dark lines behind the eyes. Eyes are pale whitish, and beak and cere are dull orangish with darker smudges.

k. Juvenile. All dark head shows dark cere, beak, and eyes. Dark breast contrasts with tawny belly, and brown back lacks white triangle. Wingtips fall short of tail tip.

l. Basic IV. Head, neck, and tail are adult-like, but with dark spots and lines; white on neck doesn't extend to the body. Body and wing coverts show white spotting. White tail and undertail coverts are variably marked blackish.

▋ OSPREY

Pandion haliaetus PAGE 102

Large, long-winged, white-bodied raptor. White head with dark eye-stripes and gull-like crooked wings are distinctive. Lack of supra-orbital ridges results in pigeon-headed look.

a. Head-on. Soars and glides with wrists elevated and wingtips depressed.

b. Adult female. All soar with wingtips and tail spread, and show unique underwing pattern. Females usually show a row of dark streaks across the breast.

c. Juvenile. All glide with wings held gull-like, with wrists pushed forward. Juveniles have paler remiges, with wide white tips, and often show a faint buffy bloom on wing linings.

d. Caribbean Osprey. *P. h. ridgwayi* has all white head with indistinct eye-stripes.

e. Juvenile. Has orangish eyes, narrow dark streaks in the crown, and white feather edges on back, wing coverts, and secondaries.

f. Adult male. Shows little or no dark breast bands. White head with dark eye-stripes is diagnostic for all. Note uniformly brown upperparts, white stripe above the wings, and long legs. All show wingtips extending beyond tail tip.

PLATE 4

GRAY-HEADED KITE AND HOOK-BILLED KITE

Two kites with paddle-shaped wings. Wingtips extend about halfway down the long tail of both kites when they are perched.

1 GRAY-HEADED KITE

Leptodon cayanensis PAGE 106

Primaries project just over halfway down the tail on perched kites.

a. Adult. Gray head, white body, and black underwing coverts are distinctive. Primaries show wide black and white banding. Secondaries are pale with indistinct narrow banding. Long black tail shows two wide white bands. Legs and feet are gray.

b. Juvenile white. Brown head shows white throat. White underparts and underwing coverts show narrow dark streaks. Primaries are narrowly banded black and white; secondaries show faint dark barring. Dark undertail shows three wide white bands.

c. Juvenile intermediate dark. Head and throat are dark. White underparts and underwing coverts show heavy wide dark streaking. Undersides of secondaries show distinct dark barring, compared to 1b.

d. Adult. Pale gray head, dark gray uppersides, paddle-shaped wings, and black tail with two pale gray bands are distinctive.

e. Juvenile intermediate light. Some juveniles show distinctive rufous hind collars and white throats with dark mesial stripe. Juveniles have pale amber eyes.

f. Head-on. They glide on bowed wings: wrists up and tips down.

g. Juvenile white. White head shows dark crown patch and dark eye-lines. White underparts are unstreaked. Note wide white hind collar.

h. Adult. Gray head, white underparts, slate upperparts, and black tail with two wide whitish bands are distinctive. Adults have gray cere, lores, face skin, and legs and feet (not shown). Adults usually have dark brown eyes, but some show gray eyes.

2 HOOK-BILLED KITE

Chondrohierax uncinatus PAGE 109

Wings pinch in at the body on soaring kites.

a. Adult male. Males have gray heads and underwing coverts, whitish barring on underparts, boldly banded black and white primaries, and dark secondaries with gray banding. Adults show two wide white bands on black undertail.

b. Adult male dark. Overall slate-gray. Black undertail shows one wide white band.

c. Juvenile male. Juveniles have white underparts and underwing coverts, both lightly barred on males. Dark undertail shows three wide white bands. Undersides of white remiges are narrowly banded.

d. Adult female head. Some kites have a huge beak but otherwise are the same in plumage and size. Note lack of supra-orbital ridge and unique lore coloration and an orangish bare spot above the whitish eyes.

e. Adult female. Females show dark crown, rufous hind collar, wide rufous-brown barring on buffy underparts, and brown upperparts. Dark uppertail usually shows a wide grayish band, with a second one sometimes visible near the base.

f. Adult male. Overall slate-gray, with white barring on underparts. Dark uppertail shows a wide grayish band.

g. Adult female. Females have dark crown, heavily barred rufous underparts, rufous underwing coverts, and narrowly banded remiges. Inner primaries show a rufous wash.

h. Adult male. Overall gray above, with a wide whitish to pale gray band on black uppertail; another white band is often visible at tail base.

i. Head-on. They glide on bowed wings: wrists up and tips down

j. Juvenile female. Juveniles have white underparts, distinctly and more heavily barred on females. Dark undertail shows three wide white bands. Juveniles have dark brown eyes.

k. Adult female dark. Overall blackish-gray, compared to the slate-gray of males. Dark uppertail shows a wide whitish band. Dark juveniles are similar, but with two white tail bands.

PLATE 5

SWALLOW-TAILED KITE AND WHITE-TAILED KITE

Two white-bodied kites with pointed wingtips.

1 SWALLOW-TAILED KITE

Elanoides forficatus PAGE 112

Unmistakable: bold black and white plumage; long, deeply forked tail; and graceful flight. Often seen in groups.

a. Adult. White head and underparts, black and white underwings, and long, deeply forked tail are distinctive. Lower belly and undertail coverts are often soiled.

b. Adult. From above, a black "mantle" of upper back and lesser and median coverts contrast with rest of blue-black upperparts. White head contrasts with dark backgrounds.

c. Head-on. Often glides with wrists held below body and wingtips.

d. Adult. Head appears pigeon-like due to lack of supra-orbital ridges above the dark brown eyes. White patches are often visible on lower back. Short legs are gray.

e. Juvenile. Like adult but with shorter tail. Narrow white edges of upperwing coverts are sometimes noticeable.

f. Juvenile. Like adult but with shorter tail and paler eyes. When seen close, a hint of buffy coloration on body and fine dark crown streaking are noticeable.

2 WHITE-TAILED KITE

Elanus leucurus PAGE 117

Falcon-shaped white kite, but gull-like in coloration. Black shoulder patches, small black carpal patches, and white tail are distinctive. Wingtips almost reach tail tip on perched kites.

a. Juvenile. Eyes are pale brown, forehead is white, and rear crown, nape, and back are gray-brown, with white tips on the latter as well as on the gray-brown upperwing coverts, black shoulder, and gray remiges.

b. Basic II. After a body molt, immatures are more adult-like, but with white tips on retained coverts and flight feathers and a narrow gray subterminal tail band. Eyes are orangish.

c. Adult female. Head, tail, and underparts are white, and eyes are scarlet. Upperparts and remiges are medium gray, with noticeable black shoulders. Females have gray crowns.

d. Adult stooping. They attack with wings held high overhead, lowering slowly at first and then accelerating quickly to pounce on prey.

e. Juvenile gliding. Somewhat adult-like, but has broad rufous wash across breast and narrow gray subterminal tail band. All show small black carpal patches.

f. Basic II from above. After a body molt, immatures are more adult-like, but with white tips on wing coverts and remiges and a narrow gray subterminal tail band. Eyes are orangish. Tail tip of all appears rounded at the corners due to shorter outer tail feathers.

g. Adult male hovering. Almost all hunting is from a hover, often with feet dangling. Males have white crowns.

h. Head-on. Soars and glides with wings usually held in a modified dihedral.

PLATE 6

DOUBLE-TOOTHED KITE AND PEARL KITE

Two small kites, both often seen perched in the open and soaring. Wingtips about two-thirds down tail tip on perched kites.

1 DOUBLE-TOOTHED KITE

Harpagus bidentatus PAGE 128

Small, accipiter-like kite. Three plumages: adult male, adult female, and juvenile. All show dark throat stripe and lack prominent supra-orbital ridges over the eyes. Ceres are dull yellow, and short legs are yellow.

a. Adult male. Blue-gray head has white throat with noticeable dark central stripe. Whitish underparts are boldly barred gray, with a rufous wash on sides of upper breast. Remiges are strongly banded, like those of accipiters, but primaries are marked bolder than secondaries. Whitish underwing coverts are rather unmarked. Fluffy white undertail coverts are often spread in flight. Tail is usually held folded in flight and shows wide, even-width black and white bands.

b. Juvenile female. Somewhat similar to adult from below, but head is brown, and underparts are streaked dark brown, with dark barring on the flanks. Females' underparts are more heavily marked compared to males.

c. Adult. Head and upperparts are blue-gray, and blackish tail shows three narrow white bands. Note the spread fluffy undertail coverts.

d. Juvenile. Somewhat similar to adult from above, but head and underparts are dark brown and undertail coverts are usually not spread.

e. Adult female. Similar to adult male from below, but breast is rufous and belly is barred rufous. Compared to 3a, wings are a bit narrower, especially the tips, tail is most often folded, and fluffy undertail coverts are spread.

f. Head-on. They soar with wings held level but glide with wingtips depressed somewhat.

g. Adult male. Blue-gray head has white throat with noticeable dark central stripe. Upperparts are blue-gray. Whitish underparts are boldly barred gray, with a rufous wash on sides of upper breast. Tail below shows wide, even-width black and dusky bands, with a narrow white band in the latter. Adult eyes are orange.

h. Adult female. Similar to adult male, but somewhat larger; breast is uniformly rufous, and belly is barred rufous.

i. Juvenile male. Brown head has white throat with noticeable dark central stripe. Upperparts are dark brown. Whitish underparts of males are lightly streaked dark brown. Tail below shows wide, even-width black and dusky bands, with a narrow white band in the latter. Juvenile eyes are yellow.

2 PEARL KITE

Gampsonyx swainsonii PAGE 115

Small, pale, falcon-like kite. Wingtips fall short of tail tip on perched kites.

a. Adult. Undersides are overall whitish, with rufous flank feathers and leg feathers and a pale cinnamon to rufous wash on leg feathers and underwing coverts. Note short, narrow, dark semi-collar. Pointed wings have dusky wingtips, and tail is square-tipped.

b. Juvenile. Almost like adult from below, but with a cinnamon wash on underparts, wing linings, and leg feathers, and lacking rufous on flanks and leg feathers.

c. Adult. Crown and upperparts are slate-gray, with obvious cinnamon wash on forehead and cheeks. Note narrow white streaks on rear inner wings (tertials) and white terminal band on inner wings formed by pale tips of secondaries.

d. Juvenile. Almost like adult from above, but with darker gray on crown and upperparts and pale edges on greater upperwing coverts.

e. Head-on. They soar and glide with wings held level.

f. Adult. When relaxed and puffed out, adults usually show lots of white feather bases on upperparts.

g. Juvenile. Almost like adult, but with white tips of primaries, faint narrow pale feather edging on wing coverts, and a cinnamon wash on underparts in fresh plumage. Hind collar is wider than adult's and lacks rufous at bottom. They lack rufous on flanks and have brown eyes and yellow legs.

h. Adult. Crown and upperparts are sooty-gray. Forehead and cheeks have a cinnamon wash. Creamy underparts show rufous flanks. Note narrow dark semi-collar on sides of upper breast and pale hind collar with narrow rufous band at bottom. Eyes are red, and legs are orange-yellow.

3 SHARP-SHINNED HAWK

Accipiter striatus PAGE 150

Single figure of soaring adult for comparison to Double-toothed Kite.

a. Adult soaring. Compare with 1a, adult Double-toothed Kite. See plate 10 for complete coverage of this hawk.

PLATE 7

SNAIL KITE AND SLENDER-BILLED KITE

Two snail-eating kites.

1 SNAIL KITE

Rostrhamus sociabilis PAGE 121

Adult males and females have different plumages. Juvenile and second-plumage kites are similar to adult females. Wingtips extend beyond tail tip on all kites when perched.

a. Adult male. Overall darkish gray, with white bases to tail and tail coverts and bright orange face skin and legs.

b. Adult female. White head shows dark eye-lines, white underparts and underwing coverts are heavily marked with dark, undersides of flight feathers are boldly barred, base of tail and undertail coverts are white, and legs are bright orange.

c. Head-on. All glide with wings heavily bowed.

d. Third plumage male. Similar to adult male but with boldly marked undersides of remiges.

e. Juvenile. Creamy head shows dark eye-lines, creamy underparts are heavily streaked, darkish underwing coverts are heavily spotted, undersides of pale flight feathers are boldly barred, base of tail and undertail coverts are white, and legs are yellow. Note that dark band on tail tip is paler than that on adult tail. Note dark patches at sides of breast.

f. Adult male. Upperwings are two-toned: darker remiges contrast with paler gray coverts. Base of dark tail and uppertail coverts are white, and legs and face skin are bright orange.

g. Adult male. Overall slaty-gray. Eyes are scarlet, and face skin and legs are bright reddish-orange. Blackish wingtips extend beyond tail tip.

h. Adult female. Whitish head shows dark eye-lines, white underparts are darkly and heavily marked. Dark brown upperparts show faint buffy feather edges, base of tail and undertail coverts are white, eyes are dark scarlet, and face skin and legs are orange.

i. Adult female (older). Older females become darker and grayer, appearing more like adult males.

j. Second plumage. Similar to juveniles, but have whiter and more heavily marked head and underparts, adult-like tail, and yellow-orange legs and face skin.

k. Juvenile. Buffy head shows dark eye-lines, buffy underparts are heavily streaked, eyes are dark brown, base of brown tail and undertail coverts are white, and face skin and legs are yellow. Note that dark band on tail tip is paler than that on adult tail.

2 SLENDER-BILLED KITE

Helicolestes hamatus PAGE 126

Overall slaty-gray, with short tail and wings. All have whitish to pale yellow eyes. Wingtips reach tip of short tail.

a. Head-on. All glide with paddle-shaped wings heavily bowed.

b. Adult. Sexes alike. Overall, unmarked darkish gray. Legs, cere, gape, and face skin are bright orange-red. Wings are pushed forward when soaring. Short tail is square-tipped.

c. Adult. Uppersides are uniformly darkish gray, except for blackish tips of outer primaries and blackish uppertail.

d. Juvenile. Like adult but with heavily barred undersides of remiges and two narrow white tail bands.

e. Adult. Overall uniformly darkish gray, with blackish primaries. Wingtips reach tip of short tail. Legs, cere, gape, and face skin are bright orange-red.

f. Juvenile. Similar to adult in being overall gray, but tail shows white bands and upperparts show brownish feather edges. Cere, gape, and face skin are bright yellow-orange to orange.

PLATE 8

PLUMBEOUS KITE AND MISSISSIPPI KITE

Two similar gray, aerial, falcon-like kites with pointed wingtips. Both show short outer primaries.

1 PLUMBEOUS KITE

Ictinia plumbea PAGE 133

Falcon-shaped dark kite. Most kites show rufous primaries. Wings are relatively longer that those of 2. Wingtips extend far beyond tail tip on perched kites, unlike 2.

a. Adult. Overall dark slate-gray with bright rufous primaries and two white tail bands.

b. Juvenile. Similar to adult, but buffy underparts are heavily streaked dark brown to slate. Rufous on primaries is less extensive. Gray-brown head shows short pale superciliary line.

c. Transition. Head and body are adult-like; remiges and rectrices are juvenile, but some show less rufous in primaries.

d. Distant adult. Rufous on primaries is obvious.

e. Juvenile. Some juveniles show whitish primaries that lack rufous.

f. Adult. Overall dark slate-gray with bright rufous primaries and two white tail bands. Note the orange legs.

g. Adult. Adults from above are slate-gray and show rufous on primaries and incomplete white tail bands.

h. Juvenile. Gray-brown head shows short, wide white superciliary line. Eyes are pale brown. Upperparts are dark brown with faint pale feather edges. Dark primaries show pale tips. Buffy underparts show heavy dark brown to slate-gray streaking. Short legs are orange-yellow.

i. Adult. Adults are overall dark slate-gray, with noticeable white tail band, and show black areas in front of the scarlet eyes and dark ceres. Short legs are bright orange.

j. Pale juvenile. Some juveniles are very whitish.

2 MISSISSIPPI KITE

Ictinia mississippiensis PAGE 130

Falcon-shaped dark kite. Flared tail tip is usually noticeable. Wingtips reach or barely extend beyond tail tip on perched kites, unlike 1.

a. Pale juvenile. Some juveniles appear overall buffy below.

b. Juvenile. Head is dark. Underparts are heavily streaked rufous-brown. Undersides of remiges show a variable amount of whitish (compare to 2d and 2f) and seldom show any rufous, unlike 1b. Underwing coverts are rufous brown. Undertail shows narrow white bands, sometimes absent (compare to 2f).

c. Basic II, late summer, autumn. Second-plumage kites appear adult-like, but have retained juvenile outer primaries and secondaries and show a variable amount of pale makings on underparts. Note new female outer rectrices.

d. Adult male. Overall gray with paler gray head. Primaries usually show some rufous. Black tail is flared near tip. Note narrow white band on trailing wing edges.

e. Paler juvenile. Some juveniles show extensive white on remiges, wider white tail bands, and paler rufous streaking on undersides and underwing coverts.

f. Adult. Adults above show paler gray head, white bars on trailing edges of upperwings, and dark flared tail. Most show no rufous on upperwings.

g. Darker juvenile. Some juveniles show little white on remiges, no white tail bands, and darker rufous-brown streaking on undersides and underwing coverts.

h. Basic II, spring, early summer. Juveniles returning in spring have gray head feathers, orangish eyes, and many gray body and covert feathers, but have retained juvenile remiges and rectrices.

i. Adult female. Adults are overall gray and show black areas in front of the scarlet eyes, dark ceres, pale gray bar on sides, and dusky front to short orange-yellow legs. Females show pale outer rectrices with dusky band on tip and white undertail coverts marked gray.

j. Adult female. Females in flight show pale undertail and whitish undertail coverts.

k. Juvenile. Darkish head shows short wide white superciliary line. Eyes are brown. Upperparts are brown. Buffy underparts show heavy rufous-brown streaking. Short legs are yellow with dusky wash on fronts. Tail tips of perched kites can appear notched.

PLATE 9

CRANE HAWK AND NORTHERN HARRIER

Two slender, long-tailed, and long-legged raptors.

1 CRANE HAWK

Geranospiza caerulescens PAGE 165

Slender, long-legged dark raptor. Wingtips barely extend beyond tail base on perched hawks. Soars occasionally. Usually hunts from perch and moves with agile grace among branches of trees, and also in harrier-like flight over open areas, especially marshes.

a. Adult. Black overall with paddle-shaped wings showing row of white spots forming a crescent on outer wings and white tail bands.

b. Adult foraging. They often reach into cavities with their long, double-jointed legs.

c. Juvenile. Similar to adult, but with some whitish on the face and breast. Eyes are yellow, and legs are orangish. Bill and cere are dark.

d. Adult. Overall black with long red legs and small head, sometimes showing a short cowl. White barring on flanks is visible on some. Eyes are red. Bill and cere are dark.

e. Juvenile. Similar to adult in flight, but with more pale markings on undersides.

f. Adult or juvenile. Both appear black above with a row of white spots forming a crescent on the outer primaries and two white tail bands.

g. Paler adult. South American adults occur in eastern Panama and are overall a paler gray.

2 NORTHERN HARRIER

Circus hudsonius PAGE 140

Distinctive long-winged and long-tailed raptor. All show white uppertail coverts and owl-like facial disk. Their slow, quartering hunting flight is distinctive. Wingtips fall short of tail tip on perched harriers.

a. Adult male. Overall whitish below, with gray head and black wingtips and band on trailing edge of wings. Underparts show a variable amount of rufous spotting.

b. Adult male. Appears darker gray above, with a variable amount of darker markings.

c. Adult female. Head is brown, eyes are yellow, and buffy underparts are heavily streaked. Note barred flanks and pale bands through dark secondaries.

d. Paler juvenile. In spring, their rufous underparts have faded to pale buffy. Pale banding on the secondaries is less obvious than that of adult females. Note large dark area formed by dark axillars, secondaries, and coverts at base of secondaries.

e. First-plumage adult male. Similar to adult male, but with brownish cast to head and upper breast and more rufous markings on underparts and underwing coverts.

f. Adult female. Appears mostly brown on uppersides, with rufous streaking on neck and markings on upperwing coverts and pale banding on remiges noticeable. Dark brown tail shows buffy bands.

g. Juvenile. Upperparts are more uniformly brown, and necks are dark. Compare to 2f. Dark brown tail shows buffy bands.

h. Head-on. Glides and hunts with wings in a strong dihedral. Hunts by slowly quartering low over the ground.

i. Juvenile male. Males have pale eyes, either pale yellow, gray, or brown.

j. Juvenile female. Females have dark brown eyes. Juveniles in fresh plumage have rufous underparts. Dark streaking on underparts, if present, is usually restricted to upper breast and flanks. Juveniles and adult female show distinct, owl-like facial disk.

k. Adult female. Buffy underparts are strongly streaked. Eyes are yellow.

l. First-plumage adult male. Like adult males, but with a variable amount of brown on the head and rufous on breast.

m. Adult male. Head and upperparts are gray, with a variable amount of darkish markings, and whitish underparts show a variable amount of rufous spotting. Eyes are lemon-yellow.

PLATE 10

TINY HAWK, WHITE-BREASTED HAWK, AND SHARP-SHINNED HAWK

Three small, bird-catching hawks. Males of all three are separably smaller than females.

1 TINY HAWK

Hieraspiza superciliosa PAGE 148

Our smallest hawk, thrush-sized, usually unobtrusive in the middle to upper stories of all but the densest tropical forests from Nicaragua south. Does not soar. Flight is direct on rapid wing beats. Primary projection is short. Short tail shows square corners on tips. Cere and lores are bright yellow, and legs are yellow-orange.

a. Adult female. Gray head shows darker crown and red-orange eyes. Upperparts are dark blue-gray, and white underparts are narrowly barred dark gray. Short tail shows black and gray bands of equal width on uppersides. Note the yellow supra-orbital ridge. Legs are relatively shorter than those of 2 and 3.

b. Adult male. Like adult female, but noticeably smaller. Folded undertail is unbanded.

c. Juvenile rufous morph (rare). Rufous head has dark crown. Rufous upperparts show some brown markings, and creamy underparts show narrow rufous barring, heavier on leg feathers. Rufous tail has narrow, dark brown bands. Juveniles have yellow to orange eyes and yellow ceres.

d. Juvenile brown morph (common). Brown head has darker crown. Upperparts are brown, and creamy underparts show narrow rufous barring, heavier on leg feathers. Short tail shows dark brown and brown bands of equal width. Juveniles have yellow to orange eyes. Cere and lores are yellow.

e. Adult. Overall gray above and barred gray below. Wings are short and broad, and remiges show bold narrow dark banding. Short tail shows black and gray bands of equal width.

f. Juvenile brown morph. Overall brown above, and barred rufous below. Wings are short and broad, and remiges show bold narrow dark banding. Short tail shows dark brown and brown bands of equal width.

2 WHITE-BREASTED HAWK

Accipiter chionogaster PAGE 154

Small hawk; resident breeder in mountain pine and oak forests from s Mexico to n Nicaragua. Long tail shows square corners on tips. Primary projection is fairly short. Cere is yellow. Legs are yellow, brighter on adults.

a. Juvenile male. Head and upperparts are dark brown. Whitish underparts show faint rufous streaking. Tail shows dark brown and grayish-brown bands of equal width.

b. Adult male. Dark crown contrasts with white throat, body, and underwing coverts. Whitish remiges have narrow dark barring. Wings are short and broad. Undertail shows black and white bands of equal width, with narrow black bands on outer feathers.

c. Adult male. Crown and upperparts are dark blue-gray, and uppertail shows black and gray bands of equal width. Note white throat. Long tail shows square corners on tips.

d. Adult male. Dark crown and upperparts contrast with white underparts. White undertail when folded shows narrow dark bands. Eyes are brownish red.

e. Juvenile male. Head and upperparts are dark brown. Note narrow pale superciliary line and whitish markings on back. Uppertail shows dark brown and gray-brown bands of equal width. Eyes are yellow.

f. Adult female. Dark crown and upperparts contrast with white underparts. Uppertail shows black and gray bands of equal width. Eyes vary from orange to red to dark brownish-red.

g. Juvenile female. Dark brown upperparts contrast with whitish underparts that are faintly streaked rufous. Leg feathers are noticeably rufous. Folded undertail shows narrow black bands.

3 SHARP-SHINNED HAWK

Accipiter striatus PAGE 150

Small hawk that is a resident breeder in the mountains of Mexico and a widespread winter visitor.

a. Adult male. Buffy body; underwing coverts are heavily barred rufous. Throat is buffy, and undertail coverts are white. Whitish remiges have narrow dark barring. Wings are short and broad. Undertail shows black and white bands of equal width, with narrow black bands on outer feathers.

b. Juvenile male. Head and upperparts are dark brown, and buffy underparts show narrow dark brown streaks. Brown head shows narrow pale superciliary line. Tail shows dark brown and grayish-brown bands of equal width.

c. Juvenile female. Brown head shows narrow pale superciliary line. Head and upperparts are dark brown, and buffy underparts show heavy rufous markings but can show narrow dark streaking, especially on 3b. Whitish folded undertail shows narrow dark brown bands. Eyes are yellow to orangish.

d. Adult male. Head and upperparts are blue-gray, and buffy underparts are heavily and narrowly blobbed and barred rufous. Eyes are orange to red. Undertail coverts are white, and whitish undertail shows narrow black bands.

e. Adult male. Head and upperparts are blue-gray. Uppertail shows black and gray bands of equal width. Note rufous throat. Long tail shows square corners on tips.

f. Adult female. Like adult male, but noticeably larger. They occasionally show a few white spots on the back.

g. Adult female suttoni. Like North American adults, but with rufous thighs and flanks.

PLATE 11

NORTHERN GOSHAWK, COOPER'S HAWK, AND BICOLORED HAWK

One large and two medium-sized accipiters. Wingtips fall way short of tail tip on perched hawks.

1 NORTHERN GOSHAWK
Accipiter gentilis PAGE 162

Our largest and most robust accipiter. Wings are proportionally long, more buteonine-like. Rare and local in mountains of w Mexico. Primary projection is longer than those of other perched accipiters.

a. Juvenile. Juveniles are heavily streaked below, including undertail coverts. Their wings are longer than other accipiters and, when soaring, often show an abrupt narrowing at the primary-secondary junction. Wide pale superciliary line is usually visible. Wingtips appear somewhat pointed in powered flight.

b. Adult. Blue-gray mantle contrasts with darker remiges. Note large head, long tapered wings, and indistinct bands on long tail.

c. Adult. Overall gray below, with bold face pattern and long wings. Inner primaries and secondaries are faintly barred. White undertail coverts are often fluffed out to the side.

d. Head-on. Glides on cupped wings. Soars with wings level.

e. Juvenile female. Head shows wide pale superciliary line and yellow eyes. Note row of buffy spots on wing coverts and dark markings on undertail coverts. Dark tail bands are wavy, with narrow white lines bordering some bands. Wingtips reach about halfway down tail.

f. Adult male. Gray adult with bold face pattern is distinctive. Eyes are dark mahogany.

2 COOPER'S HAWK
Accipiter cooperii PAGE 156

Medium-sized, large-headed accipiter. Uncommon to fairly common resident in mountains of w and s Mexico. Winter migrant throughout, rare south of Honduras.

a. Adult. Upperwings and back are blue-gray. Note shorter, rounded wings and distinct tail bands.

b. Adult male. Males have gray cheeks. Underparts and underwing coverts are boldly barred rufous. Soars with leading edge of wings held straight. Note long spatulate tail with wide white terminal band on rounded tip.

c. Juvenile male. Streaking on underparts is heaviest on breast, fainter to absent on belly and absent on undertail coverts. Note long spatulate tail with wide white terminal band on rounded tip. Usually lacks pale superciliary line. Head projects far beyond wrists on gliding hawks.

d. Adult female. Like male but larger, with rufous cheeks.

e. Head-on. Soars with wings in a slight dihedral; glides with wings level.

f. Juvenile female. Brown head often shows hackles raised and rufous cast to cheeks and usually lacks pale superciliary line. Dark brown upperparts usually show whitish spotting. Undertail coverts are unmarked. Note long tail with even banding and wide white tip. Eyes are pale yellow.

g. Juvenile. Head cutout shows pale superciliary line (uncommon).

h. Adult male. Adults have dark crowns that contrast with pale napes. Males have gray cheeks. Head appears rounded when hackles not elevated (not shown). Eyes are scarlet.

i. Adult female. Females have rufous cheeks. Underparts are boldly barred rufous. Note long tail with regular banding and wide white tip. Eyes are orangish.

3 BICOLORED HAWK
Accipiter bicolor PAGE 159

Medium-sized, inconspicuous accipiter. Uncommon in humid lowland forest. Lacks streaking or barring. Does not soar.

a. Adult. Flying adults show gray body, white underwings, rufous legs, and black tail with three narrow white bands.

b. Rufous juvenile. Crown is blackish, and underparts and underwing coverts are rufous (rare). Long blackish tail shows three narrow white bands.

c. Buffy juvenile female. Crown is blackish, upperparts are dark brown with pale feather edges, and throat, underparts, and collar are buffy. Legs are a contrasting rufous-buff.

d. Adult male. Overall gray, with blackish crown, rufous legs, white undertail coverts, and blackish tail that shows three bands: gray above and white below.

PLATE 12

ROADSIDE HAWK, PLUMBEOUS HAWK, AND SEMIPLUMBEOUS HAWK

Three small buteonines. One is common and widespread and the other two rare and local.

1 ROADSIDE HAWK

Rupornis magnirostris PAGE 186

Small buteonine with a barred belly. Common and widespread. Adult plumages vary with locality, especially in the amount of rufous in the flight feathers and the color of the head, breast, and upperparts. Wings are rather paddle-shaped. Wingtips fall short of tail tip on perched hawks. They are often noisy and soar regularly.

a. Adult *griseocauda*. Breast and head are dark gray-brown, and buffy belly is heavily banded dark rufous-brown. Buffy underwing coverts are lightly marked, and undersides of pale flight feathers show numerous narrow dark bands. Undertail shows dark and pale bands of equal width (except on outer feathers), and buffy undertail coverts are unmarked. This race usually shows little or no rufous in the flight feathers.

b. Adult *petulans*. Similar to adult *griseocauda*, but head, breast, and upperparts are overall grayer, barring on the belly is rufous, and flight feathers show a noticeable rufous wash.

c. Juvenile. Buffy underparts have dark breast streaking and dark belly barring. Buffy underwing coverts are lightly marked, and undersides of pale flight feathers show numerous narrow dark bands. Juveniles usually show no rufous in the flight feathers. Undertail shows dark and pale bands of equal width (except on outer feathers), and buffy undertail coverts are unmarked.

d. Adult *griseocauda*. Head and upperparts are gray-brown. Some show a hint of rufous in the flight feathers. Uppertail shows dark and pale bands of equal width. Note buffy uppertail coverts.

e. Juvenile. Head and upperparts are dark brown. Juveniles usually show no rufous in the flight feathers. Uppertail shows dark and pale bands of equal width, but bands are narrower and more numerous than those of adults. Note white superciliary line and buffy uppertail coverts.

f. Adult *petulans*. Similar to adult *griseocauda*, but head and upperparts are overall grayer, and flight feathers show a noticeable rufous wash.

g. Adult *petulans*. Head, breast, and upperparts are gray. Buffy belly is heavily barred rufous, and buffy leg feathers are narrowly barred rufous. Some adults show some to lots of rufous in tail band, completely so in many in Panama. Pale yellow eyes are somewhat striking, and rufous on remiges is sometimes noticeable.

h. Adult *conspectus*. Similar to adult *petulans*, but overall much paler.

i. Adult *griseocauda*. Similar to adult *petulans*, but head, breast, and upperparts are gray-brown to brown, and belly barring is rufous-brown to brown. No rufous in remiges.

j. Juvenile (pale extreme). Similar to other juveniles, but with narrower and less obvious breast streaking and belly barring. Outer tail feathers show narrower dark bands.

k. Juvenile. Head and upperparts are dark brown, with white superciliary lines and pale brown eyes. Buffy underparts show heavy dark brown breast streaking and belly barring.

2 PLUMBEOUS HAWK

Cryptoleucopteryx plumbea PAGE 167

Small, compact, dark buteonine. Rare and local in Panamanian rain forest. All ages have red eyes and orange ceres, lores, and legs. Does not soar. Short primary projection extends halfway down short tail.

a. Adult. Head and body are dark gray, upperwings are black, and underwings are white except for dark tips of remiges. Short black tail has a narrow white band.

b. Juvenile. Almost like adult, but shows whitish barring on belly, flanks, and leg feathers; white tips on undertail coverts; faint pale barring on axillars; narrow gray banding on remiges; and a second narrower white tail band near the base. Eyes are amber.

c. Adult. Head and body are dark gray, upperwings are black, and underwings are white except for dark tips of remiges. Some adults (younger?) show faint white barring on leg feathers. Short black tail has a narrow white band. Note orange cere, lores, and legs.

d. Juvenile. Almost like adult, but shows whitish barring on belly, flanks, and leg feathers; white tips on undertail coverts; and a second narrower white tail band. Eyes are amber.

3 SEMIPLUMBEOUS HAWK

Leucopternis semiplumbeus PAGE 199

Small, compact, gray and white buteonine. Rare to uncommon and local in rain forests of Panama, Costa Rica, and e Honduras. All ages have yellowish eyes and orange ceres, lores, and legs. Does not soar. Short primary projection extends barely beyond folded secondaries.

a. Adult. Head and upperparts are gray, and underparts are unmarked white. Throat often appears dusky. Underwings show white coverts and remiges, the latter with some narrow dark banding. Long black tail has a narrow white band.

b. Juvenile. Similar to adult but with whitish streaking on the head, narrow dark breast streaking, and a second white tail band.

c. Juvenile. Similar to adult but with whitish streaking on the head, narrow dark breast streaking, upperparts that appear scaly, and a second white tail band.

d. Adult. Head and upperparts are gray, and underparts are unmarked white. Long black tail has a narrow white band. Note orange cere, lores, and legs.

PLATE 13

WHITE HAWK AND BARRED HAWK

Two large, soaring buteonines. Wingtips of both fall short of tail tip on perched hawks.

1 WHITE HAWK

Pseudastur albicollis PAGE 197

Distinctive all white hawk of lowlands and foothills of tropical rain forests. Beak is dark, cere is gray, lores are black, and legs are yellow. Two subspecies differ in the amount of black markings. Juvenile plumage is similar to adults'. Wingtips fall short of tail tip on perched hawks.

a. Adult *ghiesbreghti*. Overall white with black wingtips and narrow subterminal tail band. Wide wings make tail appear short. Note wings pushed forward.

b. Juvenile *ghiesbreghti*. Like adult, but with some gray and black markings on wing coverts and secondaries and paler eyes.

c. Adult distant *costaricensis*. Wings are pushed forward. Note broad white tips to darkish remiges.

d. Juvenile *ghiesbreghti*. Like adult, but with narrower wings and grayish markings on secondaries.

e. Adult *ghiesbreghti*. Overall white with black wingtips and tail band. Eyes are dark brown.

f. Juvenile *costaricensis*. Similar to 1b, but with more heavily marked upperwing coverts, noticeable white tips on greater coverts and secondaries, and gray-barred secondaries. Tail appears longer than adults'.

g. Adult *costaricensis*. Similar to 1e, but greater secondary coverts and secondaries are dark gray with broad white tips.

h. Juvenile *costaricensis*. Almost like 1i, but remiges show gray barring and narrower white terminal band.

i. Adult *costaricensis*. Similar to 1a, but remiges below show narrow black barring and wide white terminal band. Dark subterminal tail band is wider.

j. Adult *ghiesbreghti*. Overall white above with black wingtips and tail band.

k. Adult *costaricensis*. Differs from 1j in its dark remiges and greater coverts and wider dark tail band. Tips of greater coverts form a white line across each upperwing.

l. Juvenile *costaricensis*. Almost like 1k, but with paler gray remiges above and narrower white terminal band on trailing edge of wings.

2 BARRED HAWK

Morphnarchus princeps PAGE 183

Distinctive buteonine of the mountains of Panama and Costa Rica. Eyes are dark brown.

a. Adult. Dark gray above with narrow white tail band.

b. Adult female. Females have darker gray head, breast, and upperparts. White belly and leg feathers are heavily barred dark gray. Beak is pale gray. Cere, face skin, and legs of adults are orange.

c. Adult. Soaring hawks show distinctive dark gray head and bib, rest of underparts and wide underwings pale gray, and short black tail with one narrow white band.

d. Adult male. Males have paler gray head, breast, and upperparts.

e. Juvenile. Like adults, but overall blackish-slate, with narrow white feather edges on upperparts. Beak is black, and cere, face skin, and legs are yellow.

f. Juvenile. Much like 2c, but head and bib are blackish-slate.

g. Head-on. Soars with wings in a slight dihedral.

PLATE 14

SAVANNAH HAWK AND BLACK-COLLARED HAWK

Two large rufous buteonines.

1 SAVANNAH HAWK
Buteogallus meridionalis PAGE 169
Large, slender, long-legged buteonine of open wet grasslands in Panama. Wingtips extend well beyond tail tip on perched hawks.
a. Head-on. Glides with wings cupped.
b. Adult. Distinctive. Overall rufous except for dark back and tips of primaries and black tail with white band. Underparts are barred dark brown. Note gray cheeks and upper back, amber eyes, and orange legs. They often appear rather long-necked.
c. Basic II. Second-plumage hawks are similar to adults, and also appear overall rufous, but they show face pattern, including wide pale superciliary line, and unbarred areas on underparts.
d. Adult. Overall rufous, except for wide black border of underwings and black tail with median white band. Note that long legs and feet reach tail tip.
e. Basic II. Second-plumage hawks are similar to adults, but appear overall a paler rufous, with un-barred areas on underparts.
f. Juvenile. Buffy head and underparts show dark brown markings, with dark patches on sides of upper breast. Pale underwings have a dark brown border. Pale undertail shows wide dark brown terminal band. Some show no rufous.
g. Juvenile. Overall brown and buff, with rufous only on shoulder and leg feathers. Uppertail lacks white band. Note strong face pattern, dark patches on sides of upper breast, and extremely long legs.
h. Adult. Head and upperwings are rufous, back is brown with a gray cast, and black tail shows median white band.
i. Basic II. Similar to adult, but buffy head has rufous crown and dark eye-lines.
j. Juvenile. Somewhat similar to adult, but buffy head has dark crown and eye-lines, upperwings are paler rufous, and uppertail shows pale rufous narrow banding and wide dark terminal band.

2 BLACK-COLLARED HAWK
Busarellus nigricollis PAGE 138
Distinctive, compact buteonine, usually found in lowlands near water. White head and wide black necklace are diagnostic. Adults are overall rufous; juveniles are mostly dark brown. Wingtips reach tail tip on perched birds. Eyes and cere are dark.
a. Head-on. Soars with wings held in slight dihedral.
b. Adult. White head, rufous mantle, brownish-black remiges, and rufous base to rectrices are diagnostic.
c. Transition. Molting hawks are intermediate in plumage, but always show white head, dark collar across breast, and some rufous coloration.
d. Juvenile. From above, juveniles show white heads, but brown markings in areas that are rufous in adults. Narrower wings and longer tail result in a different silhouette. Note pale bands at the base of the tail.
e. Adult. White head, black necklace, rufous body and underwing coverts, and dark brown remiges are diagnostic. Tail is banded rufous and black with black sub-terminal band. Wide wings of adults make tail appear short. Rufous inner secondaries and base of undertail show dark barring.
f. Juvenile. Somewhat like adult with white head and dark collar, but plumage is mostly brownish, with buffy streaks and mottling and often with rufous cast. Note white breast patch.
g. Darker juvenile. Some juveniles are darker and more heavily marked, lacking a rufous cast.
h. Adult. Distinctive: white head, wide black collar, dark tail tip, and otherwise rufous.
i. Juvenile. White head and dark collar are like adult, with narrower wings with pale rufous cast. Under-parts are brownish, marked buffy, usually with white breast patch.

PLATE 15

BUTEOGALLUS ADULTS

Three large to huge slate-gray buteonines with broad wings. All soar regularly.

1 COMMON BLACK HAWK

Buteogallus anthracinus PAGE 172

Smallest Buteogallus. *Two subspecies: nominate widespread and B. a. subtilis (Mangrove Black Hawk), the latter only mixing in e Panama. Usually found near bodies of water, especially coastal mangroves.*

a. Adult. Overall slate-gray, with noticeable white slashes at base of outer three primaries and grayish mottling and wide dark terminal band on the flight feathers. Legs extend only to narrow white tail band. Orange on cere extends onto base of bill. Note rounded corners of tail tip, unlike 3a, and wings pushed forward when soaring.

b. Juvenile in molt. They show a mix of new adult and old juvenile feathers.

c. Adult *subtilis*. Like nominate adult but with strong rufous wash on remiges.

d. Adult female head. Like adult male but usually with whitish area from lores to under the eyes.

e. Adult. Overall dark slate-gray with noticeable grayish on uppersides of flight feathers and narrow white tail band. Uppertail coverts are black, unlike those of 2c and 2d.

f. Adult male. Overall dark slate-gray. Orange on adults' cere extends onto base of dark bill, unlike 2e and 2f. Wingtips fall just short of tail tip, unlike those of 2f and 3d. Note grayish on folded secondaries. Legs are noticeably shorter than those of 2f. Males lack white areas under the eyes.

g. Adult *subtilis*. Like nominate adult but with strong rufous wash on remiges.

2 GREAT BLACK HAWK

Buteogallus urubitinga PAGE 176

Medium-sized Buteogallus. *Two subspecies: nominate in e Panama and B. u. ridgwayi otherwise throughout. Usually found in wet grasslands, but also in rain forest.*

a. Adult *ridgwayi*. Overall dark slate-gray, lacking white slashes at base of outer three primaries and with banding and less noticeable wide dark terminal band on the flight feathers. Wings are narrower than those of 1a and 3a, with front and back edges more parallel. Legs extend beyond tail band and almost to tip. Beak is entirely black. White barring on leg feathers is sometimes noticeable. Second narrow white tail band is usually covered by undertail coverts.

b. Adult nominate. Like B. u. ridgwayi, but the entire base of the tail is white. (Uppertail coverts are slate-gray on 1e.)

c. Adult *ridgwayi*. Overall slate-gray, with subtle grayish banding on uppersides of flight feathers and white tail band and white uppertail coverts (slate-gray on 1e).

d. Adult nominate. Like 2b, but the entire base of the tail is white, and the legs feathers lack white barring.

e. Older immature. During their second molt, they have a mix of new dark slate-gray adult and old immature feathers. Tail is adult-like.

f. Adult *ridgwayi*. Overall dark slate-gray. Bill is entirely black, and cere is yellow, unlike those of 1e. Leg feathers are narrowly barred white. Wingtips barely extend beyond folded secondaries. Note narrow grayish banding on folded secondaries. Legs are noticeably longer than those of 1f.

3 SOLITARY EAGLE

Buteogallus solitarius PAGE 180

Huge Buteogallus *with very broad triangular wings that make the tail appear short. Extremely rare and local in hilly and mountainous areas.*

a. Adult. Overall dark slate-gray, with wide dark terminal band on grayish, usually unbanded flight feathers. Wings are rather triangular. Legs almost reach tail tip. Corners of tail tip are square, unlike rounded as shown on 1a.

b. Adult. Overall more grayish on uppersides, especially the remiges, compared to 1e and 2c, with a narrow white band on blackish tail. Wide wings are rather triangular, and tail barely extends beyond wings.

c. Older immature. Immatures that are almost adult are overall dark blackish-brown with some buffy markings and adult tail and some pale areas on their heads. Wingtips extend beyond tail tip.

d. Adult. Overall dark slate-gray. Bill is entirely black. Note faint banding on folded grayish secondaries and the long legs. Wingtips extend beyond tail tip.

PLATE 16

BUTEOGALLUS IMMATURES

Three large to huge brown and buffy-streaked buteonines with broad wings. All soar regularly.

1 COMMON BLACK HAWK

Buteogallus anthracinus PAGE 172

Smallest Buteogallus. *Two subspecies: juveniles of nominate and* B. a. subtilis, *Mangrove Black Hawk, are essentially alike. Usually found near bodies of water, especially coastal mangroves.*

a. Juvenile. Buffy underparts are heavily streaked dark brown, with wide dark malar stripes extending down the sides of the neck to the breast. Buffy underwings show dark markings on the coverts, narrow dark banding on the flight feathers, and paler primaries that form a noticeable panel. White undertail has irregular dark banding.

b. Juvenile. Dark brown above, with noticeable pale head markings and pale primary panels, and white tail that has irregular narrow dark banding. Uppertail coverts are dark, unlike those of 2d and 2e.

c. Juvenile head. Some juveniles show dark throats, perhaps due to pre-formative molt.

d. Juvenile. Brown head shows buffy cheeks and wide dark malar stripes that extend onto sides of upper breast. Dark brown upperparts show some buffy mottling. Buffy underparts often fade to whitish with time and are heavily streaked. Buffy leg feathers have narrow dark barring. Primaries extend almost to tail tip, unlike those of 2f and 2g.

2 GREAT BLACK HAWK

Buteogallus urubitinga PAGE 176

Medium-sized Buteogallus. *Two subspecies: immatures of nominate and* B. u. ridgwayi *are essentially alike. Usually found in wet grasslands, but also in rain forest.*

a. Juvenile. Buffy underparts are streaked dark brown, but lack wide dark malar stripes. Buffy underwings show some dark markings on the coverts, narrow dark banding on the flight feathers, and paler primaries that form a noticeable panel. Buffy undertail has narrow dark banding.

b. Basic II. Similar to juvenile, but tail is white with irregular wide black banding, or occasionally mottling. Note new darker and more heavily barred inner primaries and a few secondaries.

c. Juvenile. Dark brown above, with noticeable pale head markings and pale primary panels, and white tail that has irregular narrow dark banding. Uppertail coverts are white, unlike those of 1b.

d. Basic II. Similar to juvenile, but tail is white with irregular wide black banding, or occasionally mottling, and appears more like 1a and 1b.

e. Basic II uppertail variant. Sometimes the black markings are with highly irregular banding or mottling.

f. Basic II. Similar to juvenile, but head is more darkly marked, and white tail has irregular black banding or mottling and appears more like 1a and 1b.

g. Juvenile. Buffy head is lightly marked, lacking wide dark malar stripes. Dark brown upperparts show some buffy mottling. Buffy underparts are streaked. Buffy leg feathers have narrow dark barring. Brown tail has narrow dark brown bands. Primaries extend barely beyond the folded secondaries, unlike those of 1d. Legs are noticeably longer than those of 1d.

3 SOLITARY EAGLE

Buteogallus solitarius PAGE 180

Huge Buteogallus. *Extremely rare and local in hilly and mountainous areas.*

a. Juvenile. Dark brown above, except for pale head with dark crown and lack of pale primary patches. Uppertail lacks distinct banding but shows darker tip.

b. Basic II. Similar to juvenile but new feathers are blacker. Shorter tail is medium gray above with a black subterminal band.

c. Juvenile. Buffy underparts are heavily streaked dark brown, with dark patches on the sides of the upper breast and uniformly dark leg feathers. Wings are rather triangularly shaped. Buffy underwings show some dark markings on the coverts, unbanded flight feathers show paler primaries, and pale undertail usually shows no banding; but both flight feathers and undertail can show some faint narrow dark banding.

d. Basic II. Similar to juvenile, but new feathers are blacker. Shorter tail is whitish below with a black subterminal band. Note dark malars that extend onto sides of upper breast.

e. Basic II. Similar to juvenile, but new feathers are blacker. New gray tail feathers are shorter with a black subterminal band. Note dark malars that extend onto sides of upper breast.

f. Juvenile. Buffy head has dark brown crown and narrow dark eye-lines. Dark brown upperparts show narrow buffy feather edges. Buffy underparts are heavily streaked dark brown, with dark patches on the upper breast. Leg feathers are uniformly dark brown. Pale brown tail usually shows no banding. Wingtips almost reach the tail tip.

PLATE 17

GRAY-LINED HAWK, GRAY HAWK, AND RED-SHOULDERED HAWK

Three medium-sized buteos. Wingtips fall short of the tail tip on all. None hover.

1 GRAY-LINED HAWK

Buteo nitidus PAGE 204

South American species resident in Panama and barely into sw Costa Rica.

a. Adult. Pale gray overall, paler than 2c. Pale underwings lack black tips on remiges, especially outer primaries. Black tail shows two white bands.

b. Juvenile. Buffy underparts are marked with dark brown blobs at sides of upper breast and belly. Note lack of dark malar stripe. Primaries are paler than secondaries. Tail has dark and pale bands of equal width, except for outer feathers.

c. Adult. Uppersides are pale gray and barred. White U on base of tail is narrow. Black uppertail often shows only one wide white band.

d. Juvenile. Head appears pale with dark crown streaks and eye-stripe, but lacks dark malar stripes. Upperwings show pale primary panels. Tail has dark and pale bands of equal width. White U at tail base is narrow.

e. Head-on. Glides with wings held level, same as 2.

f. Adult. Overall pale gray, with narrow dark barring on dorsal and ventral. Primary projection is less than 2g's. Eyes are medium brown.

g. Juvenile. Overall buffy with pale head showing dark crown streaks and dark eye-line, but lacking dark malar stripe. Underparts are marked with dark brown blobs and wide streaks. Leg feathers are unmarked, unlike 2f. Eyes are medium brown.

2 GRAY HAWK

Buteo plagiatus PAGE 201

Central American species resident from n Costa Rica to n Mexico. Northern Mexican and Arizona populations migrate south.

a. Adult. Uppersides are gray and lack barring. Usually shows two or three white tail bands. Wide white U on base of tail is distinctive.

b. Juvenile. Uppersides of wings are dark. Uppertail is pale brown with many narrow dark brown bands, the outer two progressively wider. White U on base of tail is distinctive.

c. Adult. Pale gray overall, but darker than 1a. Second narrow white tail band is usually obvious. Outer primaries are tipped black, unlike those of 1a.

d. Juvenile. Whitish underparts are narrowly streaked dark brown. Flight feathers are uniformly pale. Leading edges of wings of all are straight when soaring. Note black malar and central throat stripe.

e. Juvenile. Head is dark with white superciliary line and cheeks and dark eye-stripe and malar stripe. White uppertail coverts show dark centers.

f. Juvenile. Dark head shows white superciliary line and cheeks and dark eye-stripe and malar. Whitish underparts are marked with narrow dark brown streaks. Leg feathers are darkly barred.

g. Adult. Head and upperparts are unmarked medium gray; whitish underparts are barred gray. Primary projection is longer than that of 1f. Eye is dark brown.

3 RED-SHOULDERED HAWK

Buteo lineatus PAGE 207

Uncommon and local winter visitor to Mexico.

a. Adult. Rufous mantle, bold pied remiges, and black tail with three narrow white bands are distinctive. Note white crescent panels on primaries.

b. Adult. Rufous body and wing coverts, bold pied remiges, and black tail with three narrow white bands are distinctive. Backlit wings show white crescents on primaries.

c. Juvenile. California juveniles are similar to adults, with white crescent primary panels and black tail with three white bands.

d. Juvenile. Texas juveniles are different from adults, with dark brown upperparts, buffy crescent primary panels, white uppertail coverts, and brown tail with many dark bands.

e. Head-on. Glides with wrists up and tips down on cupped wings. Soars with wings held level.

f. Juvenile. Texas juvenile has brown upperparts and buffy underparts marked with blobs and short streaks. Long dark brown tail has narrow brown bands.

g. Adult. California adults have buffy heads, uniform rufous breasts, belly narrowly barred rufous and white, rufous shoulder, pied remiges, and black tail with white bands.

h. Adult. Texas adults are similar, but have brown heads with darker malars and white barring on rufous breast.

PLATE 18

SHORT-TAILED HAWK AND BROAD-WINGED HAWK

Two small buteos with pointed wingtips, one resident and the other a migrant and winter visitor.

1 SHORT-TAILED HAWK
Buteo brachyurus PAGE 214
Small, long-winged aerial hunter. Two color morphs: light and dark. Wingtips reach tail tip on perched hawks. Tail doesn't appear particularly short.

a. Head-on. They hunt on the wing with wings level and tips upraised and head down.

b. Adult stooping. When they spot prey, they make a long spectacular dive, similar to that of Peregrine Falcons.

c. Adult light morph soaring. Dark head and cheeks contrast with white throat. Underwings are two-toned: white coverts contrast with gray remiges, and secondaries are darker than primaries. Whitish undertail shows wide dark subterminal band and other incomplete narrower ones, but can be complete on Panamanian hawks.

d. Juvenile light morph gliding. Dark cheeks contrast with white throat; cheeks have paler streaking. Often shows narrow dark streaks on the sides of the breast. Juvenile remiges below are paler than adults', contrasting less with wing coverts. Whitish undertail shows narrow dark bands.

e. Adult dark morph soaring. Head, body, and wing coverts are dark brown. Underwings are two-toned: dark coverts contrast with pale gray remiges, and secondaries are darker than primaries. Remiges and rectrices are like those of light-morph adult. Whitish undertail shows wide dark subterminal band and other incomplete narrower ones.

f. Adult gliding from above. Dark brown upperparts sometimes show paler coverts. Wingtips appear somewhat pointed on gliding hawks.

g. Juvenile dark morph gliding. Similar to dark adult, but belly and underwing coverts are heavily spotted white. Breast is unmarked and appears as a dark bib. Whitish undertail shows narrow dark bands.

h. Adult light morph perched. Dark cheeks contrast with white throat. White underparts are unmarked. Two small rufous patches on sides of the breast are usually visible.

i. Adult dark morph perched. Head, body, and wing coverts are dark brown. Note white forehead. Wingtips reach tail tip.

2 BROAD-WINGED HAWK
Buteo platypterus PAGE 211
Small compact buteo. Dark morph is rare. Wingtips appear pointed on flying hawks and fall somewhat short of tail tip on perched hawks. Wing shape is often like a candle flame.

a. Juvenile light-morph gliding. Heavily marked juvenile. Underwings are relatively unmarked and show squarish pale primary panels when wings are backlit.

b. Adult dark morph gliding. Remiges are uniformly whitish. Tail as in light morph.

c. Juvenile dark morph gliding. Similar to adult, but has rufous streaking on underparts. Tail as in light-morph juvenile. Trailing edge of wings often appears straight when gliding.

d. Adult light-morph soaring. Unmarked to lightly marked underwings and black tail with wide white band are distinctive. Underparts are barred rufous-brown.

e. Juvenile gliding from above. Overall brown and lacks any distinctive markings.

f. Juvenile alternate tail from above. Sometimes show an accipiter-like uppertail.

g. Adult uppertail. Some adults (females?) show a second narrow white band.

h. Juvenile perched, belly band. Some show unmarked breast and resultant belly band. Note the pale brown cheeks, dark eye-lines and malars, pale superciliary line, and short legs.

i. Juvenile perched, unmarked underparts. A few have unmarked underparts, but show facial markings. Wingtips fall short to tail tip. Eyes are pale brown.

j. Adult perched. Head is dark brown, except for darker malars and pale throat. Underparts are barred rufous-brown, sometimes forming a bib. Eyes are dark brown.

k. Head-on. Glides with wings held level or slightly depressed.

1a

1b

1c

1d

1e

1f

1g

1h

1i

2a

2b

2c

2d

2e

2f

2g

2h

2i

2j

2k

PLATE 19

SWAINSON'S HAWK

SWAINSON'S HAWK
Buteo swainsoni **PAGE 218**

Slender, long- and narrow-winged buteo that has pointed wingtips. White undertail coverts of dark hawks and two-toned underwings are distinctive. Wingtips reach or barely exceed the tail tip on perched hawks.

a. Adult light-morph male. Adults show two-toned underwings: white coverts contrast with dark gray flight feathers. Adult males usually have rufous breasts and gray sides of the head. All show pointed tips of long slender wings.

b. Adult light-morph female. Adult females usually have dark brown breast and often show barring on the rest of the underparts.

c. Adult light morph, pale variant. Some paler adults show a white split in the center of the dark breast.

d. Adult dark morph. White undertail coverts contrast with uniformly dark belly. Underwing coverts vary from whitish to rufous (shown) to occasionally dark.

e. Basic II light. Second-plumage hawks in spring are adult-like, but their breasts are not uniformly dark, and new remiges are darker than retained ones.

f. Adult rufous-morph male. Adult males have uniformly rufous underparts and gray sides of the head. Note the white undertail coverts.

g. Adult rufous-morph female. Adult females have dark brown breasts, rufous bellies, and brown sides of the head. Note the white undertail coverts.

h. Adult dark morph, dark underwings. Some dark adults have dark underwing coverts; a few have dark undertail coverts (not shown).

i. Basic II light, molting. Second-plumage hawks in autumn are juvenile-like, but with some darker remiges and new rectrices with wide dark subterminal band.

j. Juvenile light morph, fresh darker. Juveniles in fresh plumage have a buffy wash on underparts and underwing coverts. Note the two dark spots on the sides of the upper breast. Remiges are paler than adults', yet contrast with paler coverts.

k. Juvenile light morph, faded plumage. Juveniles returning in the spring often have whitish heads. Note the two dark spots on the sides of the breast.

l. Juvenile dark morph. Dark juveniles have heavily streaked underparts. Note the pale undertail coverts that contrast with the darker belly. Underwings are two-toned.

m. Juvenile light morph, fresh paler. Some juveniles are very pale.

n. Adult from above. Brown back and upperwing coverts contrast with dark brown remiges, and uppertail is grayish. Most show a white U on uppertail coverts.

o. Juvenile from above. Back and upperwing coverts are dark brown with pale feather edging, and uppertail is brownish. Most show pale mottling in the scapulars and a white U on uppertail coverts.

p. Head-on. They soar and glide with wings held in a dihedral.

q. Adult from above, darker. Some darker adults don't show the white U on the uppertail coverts.

r. Adult light-morph male. Adult males usually have rufous breasts and gray sides of the head. Note the white throat. Wingtips extend beyond the tail tip on perched adults.

s. Juvenile light morph female. Juveniles have a patterned head and show two dark brown patches on the sides of the upper breast. Brown upperparts show pale feather edging. Wingtips reach the tail tip on perched juveniles.

PLATE 20

WHITE-TAILED HAWK

WHITE-TAILED HAWK

Geranoaetus albicaudatus PAGE 193

Long- and broad-winged buteonine with pointed wingtips and three immature plumages. Adults and juveniles appear different. Wingtips far exceed the tail tip on perched hawks.

a. Adult male. Adults have white underparts and underwing coverts and white tails with a wide black subterminal band. Secondaries are paler than primaries. Adult male has pale gray head, clear white throat, and faint dark barring on the belly and flanks.

b. Basic III male. Somewhat similar to adult but always with slate head, throat, and upperparts and dark markings on underwing coverts and flanks. Males tend to have some rufous markings.

c. Basic II male. Similar to juveniles but with a larger, ovalish white breast patch, wider wings, and shorter tail with narrow dusky subterminal band. Males tend to have some rufous on underwing coverts and belly.

d. Juvenile male, paler. Some males have whitish underparts and underwing coverts, often showing a dark belly band like a Red-tailed Hawk.

e. Adult above. Adults show gray head and uppersides, with noticeable rufous patches on the front of upperwings, darker remiges, and white lower back and tail; tail has a wide black subterminal band.

f. Basic II tail cutout. Many replace from one to four tail feathers a second time, usually the central ones. New feathers are more adult-like.

g. Basic II above. Differs from juvenile by rufous patches on the front of the upperwings, wider secondaries, and shorter grayish tail with a narrow dusky subterminal band.

h. Juvenile above. Overall dark, with a grayish tail that has numerous faint narrow bands. Wings are narrower and tail is longer than those of older hawks. Note the white U on the uppertail coverts.

i. Juvenile female. Overall dark, with a noticeable white slash on the breast. Wings are narrower and tail is longer than those of older hawks. Underwings are somewhat two-toned: dark coverts contrast with paler remiges. Tail appears pale below; banding usually isn't noticeable. Note dark throat and variable whitish patches on head.

j. Hawks on the ground. They often occur in small groups after grass fires, on recently plowed fields, and on harvested sugarcane fields, both flying around and perching.

k. Head-on silhouette. Always flies with wings in a strong dihedral.

l. Juvenile female, darker. Darkest juveniles, usually females, don't show white breast slash or pale face marks.

m. Adult female. Adults have gray head and upperparts, white throat, and large rufous patch on upperwings. Wingtips far exceed the tail tip. Females have a medium gray head and upperparts and sometimes show gray streaking in the throat and narrow dark barring on the flanks.

n. Juvenile female. Juveniles are dark overall, with noticeable pale patches on the cheek and a white slash on the breast. They usually don't show rufous on the upperwing coverts. Wingtips extend beyond the tail tip.

o. Basic II. Similar to juvenile, but cheeks are all dark, white breast patch is larger and ovalish, upperwings show a rufous patch, tail is medium gray with a narrow dusky subterminal band, and wingtips far exceed the tail tip.

p. Basic III. Similar to adults, but head and upperparts are darker slate, throat is dark, and barring on flanks is heavier.

PLATE 21

RED-TAILED HAWK

Large buteo of highlands and mountains. Three subspecies occur: *B. j. costaricensis* is widespread in the mountains from s Mexico to n Panama, including Belize. *B. j. fuertesi* is resident in parts of n Mexico, and *B. j. calurus* is resident on Baja California and a winter visitor throughout, south to Nicaragua. *B. j. fuertesi* occurs only in light morph, but the other two have all three morphs.

RED-TAILED HAWK

Buteo jamaicensis PAGE 225

All except dark morph show large dark patagial marks on the leading edge of the underwings. Adults always have rufous tail with narrow dark subterminal band and sometimes other narrow dark bands. Wingtips reach or almost reach (adults) or fall somewhat short of (juveniles) tail tip on perched hawks.

a. Adult *costaricensis*. Adults are distinguished from other subspecies by the reddish wash on the belly, flanks, leg feathers, undertail coverts, and underwing coverts. They show a few dark streaks on the belly and a whitish throat and breast. Adults have wide wings that show a wider dark terminal band and dark eyes.

b. Head-on. All glide with wings held level or nearly so, but soar with wings raised into a slight dihedral.

c. Juvenile rufous morph. Buffy underparts and underwing coverts are heavily streaked dark brown, usually narrower on the breast. The dark patagial marks are noticeable. Juvenile undertail is pale brown with narrow dark brown bands. This could be a juvenile of either *B. j. costaricensis* or *B. j. calurus*. Juveniles, compared to adults, have narrower wings with a narrower dark terminal band and yellow eyes.

d. Adult *costaricensis*. They have rather uniformly dark head and upperparts, with little or no pale markings on the scapulars. Rufous tail usually shows a narrow dark subterminal band.

e. Adult *calurus*. Compared to 1d, they have paler upperparts and paler brown crown and cheeks and show buffy markings on the scapulars.

f. Juvenile *costaricensis*. Undersides are very whitish, including the throat. Belly band and tail bands are rather indistinct. Note the dark patagial marks of all Red-tails.

g. Juvenile *costaricensis*. Similar to juvenile *calurus* and not always separable. But they are usually overall a paler brown with a white throat. Brown tail shows many narrow dark brown bands and sometimes has a rufous cast. Juveniles show two-toned upperwings: pale primaries and coverts contrast with darker secondaries and coverts.

h. Juvenile *calurus*. Juveniles have paler dark brown upperparts that lack the warm tones of adults. Brown head sometimes shows buffy cheeks and superciliary lines, but can be completely dark. Back and upperwing coverts often show whitish spotting, and uppertail coverts are also spotted white. Brown tail shows many narrow dark brown bands and sometimes has a rufous cast.

i. Juvenile dark morph. Overall dark brown, with some rufous streaking on underparts. Tips of outer primaries are uniformly dark. Dark tail bands are usually wider than those of other morphs. Dark-morph *costaricensis* and *calurus* are essentially the same. Last dark band is almost on tail tip and lacks spike in center.

j. Adult dark morph. Dark-morph *costaricensis* and *calurus* are essentially the same. Head, underparts, and underwing coverts are uniformly dark brown. Dark undertail coverts often show rufous barring. Tail can be like those of light-morph adult.

k. Adult *costaricensis*. They have rather uniformly dark head and upperparts and show few or no pale markings on the scapulars. Rufous tails usually show a narrow dark subterminal band.

l. Adult *calurus*. Compared to 1j, they have paler upperparts and paler brown crown and cheeks, with contrasting darker throat, and show buffy markings on the scapulars.

m. Adult *costaricensis*. Head and upperparts are a uniform dark brown. Some adults, especially in southern Mexico, show reddish barring on flanks and leg feathers.

n. Adult dark morph. Dark-morph *costaricensis* and *calurus* are essentially alike. Head, body, and wing coverts are uniformly dark brown. Leg feathers often show rufous barring. Tail can be heavily banded.

o. Adult *calurus*. Head is brown with darker brown throat and cheeks. Dark upperparts show buffy markings on the scapulars. Rufous-buff underparts show dark brown markings, usually heavier on the belly, and lack the reddish wash on the belly shown by 1m.

p. Adult *fuertesi*. Similar to 1o, but with more white in the throat and few or no dark markings on creamy underparts

q. Adult faded *costaricensis*. Some adults show reddish wash on belly and leg feathers that has faded to pale buff.

r. Adult rufous morph. Similar to 1m, but with rufous breast, leg feathers, and undertail coverts. Note the wide dark belly band. Rufous-morph *costaricensis* and *calurus* are essentially alike.

s. Juvenile *costaricensis*. Brown head shows buffy cheeks and superciliary lines. Belly band is usually fairly narrow, and leg feathers are barred rufous. Back and upperwing coverts often show whitish spotting. Brown tail shows many narrow dark brown bands.

t. Juvenile *calurus*. Brown head sometimes shows buffy cheeks and superciliary lines, but can be completely dark. Belly band is usually quite wide, and leg feathers are barred dark brown. Back and upperwing coverts often show whitish spotting. Brown tail shows many narrow dark brown bands.

PLATE 22

ZONE-TAILED HAWK, ROUGH-LEGGED HAWK, AND FERRUGINOUS HAWK

1 ZONE-TAILED HAWK

Buteo albonotatus PAGE 222

Mimics Turkey Vulture. Occurs throughout; breeds in hills and mountains.

a. Juvenile. Overall black, with narrowly barred silvery remiges. They usually show white spotting on the body. Pale undertail shows dark bands when seen well.

b. Adult. Overall black, with silvery remiges. Like Turkey Vulture but with yellow cere, wide black band on hind wing, and white tail band. Rocks like Turkey Vulture in flight.

c. Adult with Turkey Vultures. They often join flocks of Turkey Vultures, flying low in the thermal where they appear farther away due to their smaller size.

d. Adult. Appears all dark from above, except for upperwings, which show somewhat paler primaries, but not as obvious as those of juvenile Red-tailed Hawks. Tail bands are not obvious unless tail is widely fanned.

e. Juvenile. All dark above, except for upperwings, which show somewhat paler primaries.

f. Head-on. Soars and glides with wings in a strong dihedral, often rocking.

g. Adult. Overall black, with wide white band on undertail. Note yellow cere, all black bill, and wingtips extending beyond tail tip.

2 ROUGH-LEGGED HAWK

Buteo lagopus PAGE 237

Occurs in winter in nc Mexico. Tarsi are loosely feathered to toes. Wingtips reach (juveniles) or extend beyond (adults) tail tip on perched hawks.

a. Adult dark. Body and wing coverts are dark blackish-brown. Sexes are alike. Black carpals contrast with wing linings. Some show white undertails with black terminal band. Some males are overall black.

b. Juvenile. Brown above with white primary panels, but shows dark primary coverts, and white uppertail coverts and base of tail.

c. Juvenile. White underparts show wide dark brown belly band. White underwings show bold black carpal patches. White undertail shows dusky band on tip.

d. Head-on. Soars with wings held in a dihedral; glides with a modified dihedral.

e. Adult female. Adults have dark eyes. Females have dark flanks. Brown tail has white base and narrow dark subterminal band.

f. Adult male. Males have pale barring on dark flanks (also white flanks with or without dark bars). Back shows gray barring, and breast is usually more heavily marked than females'. White-based gray tail usually with multiple dark bands, subterminal widest.

g. Juvenile. Dark juveniles are overall dark brown, with pale eyes. Some show some pale areas on the head. Gape doesn't extend to below eye.

3 FERRUGINOUS HAWK

Buteo regalis PAGE 234

Occurs in winter in n Mexico. Tarsi are loosely feathered to toes. Gape is large. Wingtips fall short of tail tip on perched hawks.

a. Adult dark. Body and wing coverts are dark brown, with rufous patch and white streaking on the breast. Outer primaries show narrow dark tips, trailing edges of wings lack bold dark borders, and undertail is unbanded. Note white wrist commas.

b. Head-on. Soars with wings in a strong dihedral.

c. Juvenile. Brown above with white primary panels, but dark coverts, white uppertail coverts and base of tail. Differs from 2b by narrower dark on wingtips.

d. Adult. Overall white below with dark V on belly. Note narrow dark tips on outer primaries. Undertail is unbanded.

e. Juvenile. Overall white below. Note dark line at base of wings and narrow dark tips on outer primaries. White undertail has dusky band on tip and can show faint dark banding.

f. Juvenile. Dark juveniles are overall dark brown, with pale eyes. Gape extends to below eye. Note gray cast to primaries.

g. Adult. Adults have whitish head with narrow dark eye-line and lack dark malar stripes. Upperparts are rufous. Leg feathers are rufous. White-based tail is gray or pale rufous or a combination of the two.

PLATE 23

GOLDEN EAGLE AND HARRIS'S HAWK

A large dark eagle and a dark buteonine that shows rufous in its plumage.

1 GOLDEN EAGLE

Aquila chrysaetos PAGE 245

Large, long-winged dark eagle. Golden crown and nape are distinctive. Tarsi are completely feathered. Bill and cere are tricolored. Head protrudes less than half the tail length on flying eagles. Wingtips almost reach tail tip.

a. Juvenile. Overall dark brown, with variably sized white wing panels. Tail is two-toned: white base and dark tip, with an even border between them.

b. Juvenile wing cutout. Some lack white wing panels.

c. Juvenile wing cutout. Some have extensive white wing panels.

d. Adult. Dark brown with golden crown and nape and wide, tawny-buff bar on wing coverts. Bill has dark tip and horn-colored base; cere is yellow. Wingtips almost reach tail tip.

e. Basic II. Second-plumage eagles are like juveniles, but tail shows gray areas on dark tip, and tawny-buff wing bar is noticeable.

f. Older immature. Appears adult-like, but with some longer juvenile secondaries and white on bases of some tail and flight feathers.

g. Adult. Appears overall dark; in good light, pale gray mottling and dark band on trailing edge of remiges and rectrices are often noticeable. Note the thin tawny line on the forewings, small rufous patch at the wrists, and tawny undertail coverts.

h. Juvenile from above. Overall dark brown, with two-toned uppertail, golden crown and nape, and often small white wing panels. Lacks tawny bars on upperwings.

i. Older immature from above. Adult-like but with white areas in the rectrices.

j. Adult from above. Appears brown with golden crown and nape and wide tawny bars on upperwings. Head extends less than half tail length.

k. Head-on. Glides with wings held in a dihedral.

2 HARRIS'S HAWK

Parabuteo unicinctus PAGE 190

Long-tailed, dark buteonine. Paddle-shaped wings, rufous wing coverts, and long tail are distinctive. Wingtips reach only halfway to tail tip.

a. Adult. Overall dark brown, except for rufous wing coverts and white undertail coverts and tail base. Long tail shows wide white tip.

b. Basic II wing cutout. First-plumage adults usually have paler primaries, with retained juvenile outer ones.

c. Juvenile. Similar to adult, but shows dark narrow barring on undersides of pale remiges, white streaking on underparts, and narrow white tip of narrowly banded tail.

d. Adult from above. Overall dark brown, except for rufous wing coverts and white uppertail coverts and tail base. Long tail shows wide white tip.

e. Juvenile from above. Similar to adult, but with less white on base and tip of tail.

f. Juvenile female. Females are usually less heavily streaked white, compared to 2g, appearing dark-breasted. Some juveniles have leg feathers that are heavily barred rufous.

g. Juvenile male. Males appear paler than 2f. Wingtips reach only halfway to tail tip. Some juveniles have leg feathers that are lightly barred rufous.

h. Adult perched. Overall dark brown, except for rufous wing coverts and leg feathers and white undertail coverts and tail base. Long tail shows wide white tip.

i. Head-on. Glides on bowed wings, with wrists up and tips down.

PLATE 24

CRESTED EAGLE AND HARPY EAGLE

Two somewhat similar crested forest eagles: one is large, and the other is huge. Both have unfeathered lower tarsi and show almost no primary projection beyond the folded secondaries, and neither soars.

1 CRESTED EAGLE

Morphnus guianensis PAGE 240

Large, slender, long-tailed forest eagle that shows a single long crest of two feathers. Dark eyes, lores, and cere and beak form a shallow V on the front of the face, best seen head-on. The tail is proportionally longer than that of 2. Rare and seldom seen due to secretive behavior.

a. Juvenile. Head is whitish, and back and secondary coverts are pale gray and somewhat mottled. Exposed secondaries are darkly barred. Long pale gray tail shows faint dark banding. Note single long crest and dark V on face.

b. Adult. Head and breast are gray, and white belly shows narrow rufous barring. Back and wing coverts are blackish. Long black tail shows two wide white bands below. Long central crest feathers are mostly dark. Legs are longer and slimmer than those of 2.

c. Juvenile. Head and upperwing coverts are whitish, and flight feathers are darkly barred. Long pale gray tail shows faint dark banding.

d. Adult. Pale gray head contrasts with blackish upperparts, and long black tail shows grayish bands. Note dark crest feathers on nape.

e. Older immature. Similar to juvenile, but with grayer head and breast, dark tips to long central crest feathers, dark back, gray mottled upperwing coverts, and more adult-like tail feathers.

f. Juvenile. Head and underparts are whitish, and long, pale gray tail shows faint dark banding. Flight feathers show dark barring.

g. Adult. Gray head and breast contrast with white belly. White underwing coverts distinguish adults from 2a. Flight feathers are boldly barred black and white. Long black tail shows wide white bands on undersides. They do not soar.

h. Dark-morph adult. Head, upperparts, and breast are blackish, with bold black barring on the white belly. Black face markings, leg color, and tail are the same as the light-morph adult's.

2 HARPY EAGLE

Harpia harpyja PAGE 242

Huge, bulky, and powerful forest eagle that shows two crests, one on each side. Rare, but not particularly shy. Wings of flying eagles are paddle-shaped, and wingtips barely extend beyond folded secondaries on perched eagles. Legs are yellow. They do not soar.

a. Adult. Black breast band that contrasts with gray head and whitish belly and black underwing coverts are distinctive. Flight feathers are boldly barred black and white. Long black tail shows wide white bands on undersides.

b. Juvenile. Head and underparts are whitish, with broad pale gray band across chest, and long pale gray tail shows narrow dark banding. Tips of outer primaries are dark, and the rest of the white remiges are narrowly banded.

c. Juvenile. Head is white, and white underparts show a pale gray band across the breast. Pale gray back and secondary coverts are somewhat mottled. Long white crest often appears as a cowl. Exposed secondaries are dark. Long, pale gray tail shows narrow distinct dark banding. Note the dark brown eyes and gray lores.

d. Adult from above. Gray head contrasts with blackish upperparts, and long black tail shows grayish bands. Dark flight feathers are narrowly banded.

e. Juvenile from above. Head is white, upperwing coverts are pale gray, and dark flight feathers are narrowly banded. Long pale gray tail shows narrow dark banding. Eyes appear dark.

f. Younger immature. Similar to juvenile, but head is pale gray, whitish underparts show wide gray breast band, and crest shows dusky tips. Note uneven mix of new darker feathers in gray breast band, back, and upperwings.

g. Older immature. Similar to adult, often with irregular mix of paler retained immature feathers in upperwing coverts.

h. Adult perched. Gray head shows black crest feathers. Black breast band is distinctive. Upperparts are black. Belly is white, and white leg feathers are narrowly barred dark gray. Note short primary projection and two wide, pale gray tail bands. Legs are shorter and thicker than those of 1h. Eyes are pale gray.

PLATE 25

BLACK AND WHITE EAGLE, BLACK HAWK-EAGLE, AND ORNATE HAWK-EAGLE

Three small eagles with feathered tarsi.

① BLACK AND WHITE EAGLE

Spizaetus melanoleucus PAGE 253

Small black and white aerial eagle. Wingtips are rather pointed on flying eagles (wings are not paddle-shaped, but are more buteonine) and fall short of tail tip on perched eagles. Eyes are yellow, and cere is bright orange.

a. Adult female. White head shows black crown, lores, and eye-rings, and orange cere and gape. Upperparts are black, and underparts, including leg feathers, are unmarked white. Note white at bend of wings ("shoulder").

b. Juvenile male. Very similar to adult, but with white specks in the black crown, brownish cast and pale spotting on upperparts, and more and narrower dark tail bands.

c. Head-on. All show white leading edge of inner wings. Soars and glides with wings held level.

d. Adult. From above, white head with black crown and black upperparts are distinctive. Uppertail shows black and paler gray bands of equal width.

e. Juvenile. Very similar to adult, but with white specks in the black crown, brownish cast and pale spotting on upperparts, and more and narrower dark tail bands.

f. Adult male. Overall white below, with black and orange areas on the head and lightly banded remiges. Note subterminal band noticeably broader in adult wings and tails.

g. Juvenile female. Very similar to adult, but with more and narrower dark tail bands.

② BLACK HAWK-EAGLE

Spizaetus tyrannus PAGE 248

Small black forest eagle. Flying eagles show paddle-shaped wings and long tails. Perched eagles show primaries barely extending beyond folded secondaries.

a. Juvenile. Soars and glides with wings held level. Juveniles show white face pattern.

b. Juvenile. Juveniles above are overall brown, with whitish face pattern and narrow blackish bands on long brown tail.

c. Adult. From above, adults are black with whitish spotting on outer primaries, whitish barring on uppertail coverts, and a long tail with black and pale gray bands of equal width.

d. Adult male. From below, black head and body, whitish underwing coverts with fine blackish markings, whitish remiges with bold black banding, and whitish legs with narrow black barring are distinctive. Tail shows black and pale gray bands of equal width.

e. Juvenile female. From below, juvenile is somewhat like adult, but overall browner, with a white throat and narrower and more numerous dark tail bands. Note the dark auricular patch.

f. Paler juvenile. Juveniles show brownish head with whitish superciliary lines and throat and dark auriculars, narrow dark streaks on pale breast, dark barring on the pale belly, narrow dark barring on the leg feathers, and pale and dark brown bands on the long tail. They usually lack the dark auricular patch.

g. Darker juvenile female. Similar to paler juvenile, but darker overall, with a dark auricular patch.

h. Basic II. Second-plumage eagles are adult-like, but with more whitish areas on face and in body plumage.

i. Adult male. Overall black, with white in raised cowl and bold whitish barring on the leg feathers and undertail coverts. Long tail has black and mottled gray bands of equal width. Eyes are orange.

③ ORNATE HAWK-EAGLE

Spizaetus ornatus PAGE 250

Small, long-crested eagle of forest and semi-open areas. Adults have distinctive plumage, but juveniles are overall brown and white. Flying eagles show somewhat paddle-shaped wings and long tails. Perched eagles show primaries barely extending beyond folded secondaries.

a. Juvenile. All soar and glide with wings held level. Juveniles show whitish head and brown upperparts and dark leading edge of wings.

b. Juvenile. Completely white head and brown upperparts are distinctive. Note long tail with dark brown and pale brown bands of equal width.

c. Adult. Adults above are blackish-brown with black crown and orange cheeks and nape. Note long tail with blackish and gray-brown bands of equal width.

d. Adult male. Soaring adults show orange cheeks, white throat and upper breast, and boldly barred belly. Whitish underwings show small blackish marking on coverts and narrow bold banding on remiges. Long whitish undertail shows black bands.

e. Juvenile female. Soaring juveniles show white head and underparts, with narrow black barring on flanks and leg feathers. Whitish underwings show sparse small blackish marking on coverts and narrow banding on remiges.

f. Juvenile male. Distinctive, with white head, crest, and underparts, bold narrow black barring on flanks and leg feathers, and blackish-brown upperparts. Long whitish undertail shows narrow black bands.

g. Basic II female. Second-plumage eagles are intermediate between juvenile and adult, but start to show the adult pattern.

h. Adult female. Distinctive, with black crown and long crest; white throat with narrow black malar stripes and orangish cheeks, hind neck, and upper breast; blackish upperparts; and belly and leg feathers boldly barred black and white. Long whitish undertail shows black bands.

PLATE 26

LARGE CARACARAS

Two unusual falconids.

1 CRESTED CARACARA

Caracara plancus PAGE 269

Three age-related plumages. They are large-headed, long-winged, and long-legged and have a large, pale gray beak. All show large pale primary panels and a pale tail with a wide dark terminal band. The long head and neck of flying caracaras is distinctive. Wingtips fall just short of the tail tip on perched caracaras. Long pale tail shows a wide dusky terminal band.

a. Juvenile. Juveniles are mostly dark brown and buffy. Underparts are streaked, face skin is pink, eyes are dark brown, and legs are dull grayish-yellow to yellow.

b. Basic II. Very similar to adult but more brownish-black. Barring on breast and upper back is less well defined. Face skin is orange. Brownish-black belly can show pale barring. Terminal tail band darker than juveniles'.

c. Head-on. They glide on cupped wings.

d. Adult. Adults are black and white. White breast is barred black, and belly is uniformly black, forming a wide belly band. Wing edges often appear parallel. Face skin is bright reddish-orange, and legs are yellow. Long pale tail shows a wide black terminal band.

e. Juvenile. Juveniles above are brown, with dark streaking on upper back.

f. Basic II. Very similar to adult, but more brownish-black. Barring on upper back is less well-defined. Brownish-black crown shows narrow rufous streaking. Face skin is orange.

g. Adult. Adults are black and white, with large white primary panels. White upper back is barred black.

h. Adult. Adults are black and white. White breast and upper back are barred black, and belly is uniformly black. Crown is uniformly black. Face skin is bright yellow, but can change to red quickly when excited. Legs are bright yellow, and eyes are amber.

i. Adult. They often toss their heads back onto their backs when giving their rattling call.

j. Basic II. Very similar to adult but overall more brownish-black. Barring on breast and upper back is less well defined. Face skin is orange, and legs are yellow. Brownish-black crown shows narrow rufous streaking. Brownish-black belly can show pale barring. All are comfortable walking on the ground.

k. Juvenile. Juveniles are mostly dark brown and buffy. Underparts and upper back are streaked, face skin is pink, eyes are dark brown, and legs are dull grayish-yellow to yellow. Wingtips do not reach the tail tip on perched caracaras. In fresh plumage, upperparts show pale feather edges.

2 RED-THROATED CARACARA

Ibycter americanus PAGE 267

Distinctive black and white raptors, with a large red throat patch (adults). They are social and communal and specialize on bee and wasp nests. Wingtips do not reach the tail tip on perched caracaras. Often quite vocal.

a. Juvenile. Much like adult, but black plumage less glossy; face skin is grayish with some red highlights, and legs are orangish-yellow.

b. Adult. Overall glossy black, except for white belly, leg feathers, and undertail coverts and large red throat patch. Face skin, large throat patch, and legs are reddish. Long black tail has a rounded tip.

c. Adult gliding. Overall glossy black above.

d. Adult. Overall glossy black, except for white belly, leg feathers, and undertail coverts. Face skin, large throat patch, and legs are reddish, cere and base of mandible are pale blue-gray, eyes are red to reddish-brown, and beak is yellow.

e. Juvenile. Much like adult, but black plumage less glossy; face skin is grayish with some red highlights, eyes are brown, and legs are paler red.

PLATE 27

LAUGHING FALCON AND YELLOW-HEADED CARACARA

Two unusual falconids.

1 LAUGHING FALCON

Herpetotheres cachinnans PAGE 257

Medium-sized, large-headed, paddle-winged falconid. Blackish "mask" with connecting dark strap across the nape is diagnostic. Eyes are dark brown, and cere varies from yellow to orange-yellow. Wingtips extend barely beyond the secondaries on perched falcons. They prey primarily on snakes.

a. Head-on. They glide with wings cupped.

b. Juvenile. Almost like adult, but with faint narrow crown streaking, faint narrow streaks on underparts, and narrower dark subterminal band on remiges.

c. Adult. Overall creamy below with bold black face mask, short, paddle-shaped wings with narrowly barred remiges, and long and boldly banded tail. Some show faint markings on underwing coverts.

d. Adult. Large creamy head shows large dark brown face mask. Underparts and broad lower nape band are creamy-buff. Upperparts are dark brown. Some individuals in South America show narrow pale scapulars.

e. Juvenile. Almost like adult, but with faint narrow crown streaking, faint narrow streaks on underparts, narrower rufous feather edges on upperparts, and buffy tail bands.

f. Adult above. Creamy head shows large dark face patch and narrow dark band across the nape. Upperparts are dark brown with large creamy primary panels. Creamy uppertail coverts form a U. Long dark brown tail shows whitish bands.

2 YELLOW-HEADED CARACARA

Milvago chimachima PAGE 273

Small slender, long-tailed caracara. Adult and juvenile plumages differ. Wingtips fall somewhat short of the tail tip on perched caracaras.

a. Juvenile. Overall dark and streaky. Head shows streaked crown and throat and wide dark eye-line. Creamy underparts are heavily streaked. Long tail shows narrow dark brown and cream banding.

b. Adult. Head, underparts, and underwing coverts are unmarked creamy-buff. Secondaries are dark brown and lightly marked. Primaries have creamy bases forming pale wing panels and wide dark wingtips. Creamy tail shows some narrow dark bands and a wide dark subterminal band. Note the narrow, dark eye-stripe.

c. Juvenile above. Pale head and neck are narrowly streaked. Dark upperwings show pale primary patches. Uppertail coverts are white, and long tail shows narrow dark brown and cream banding. Note the dark auricular patch.

d. Adult above. Creamy head shows narrow dark lines behind the eyes. Dark upperwings show pale primary patches. Uppertail coverts are whitish, and creamy tail shows some narrow dark bands and a wide dark subterminal band.

e. Juvenile. Overall dark and streaky. Head shows streaked crown and throat, dark auriculars, and dark eye-line extending onto hind neck. Dark brown upperparts are unmarked or with some pale feather edges. Creamy underparts are heavily streaked. Long tail shows narrow dark brown and cream banding. Beak is yellowish. Eyes are dark brown, and cere and eye-ring are pale yellow to pinkish. Legs are dull yellow.

f. Adult. Head, underparts, and underwing coverts are unmarked creamy-buff. Back and upperwing coverts are dark brown. Creamy tail shows some narrow dark bands and a wide dark subterminal band. Eyes are medium brown to dark hazel. Beak is pale blue. Cere and eye-ring are orange yellow to pinkish. Legs are dull yellow. All frequently walk on the ground.

PLATE 28

FOREST-FALCONS

Three similar, usually secretive forest raptors. All have long legs and long tails with graduated tips and, when perched, show short primary projection, barely beyond secondaries.

1 BARRED FOREST-FALCON

Micrastur ruficollis PAGE 259

Smallest. Usually found in lower levels of rain forest.

a. Adult male. Adults have gray head and upperparts, and white underparts with narrow gray barring (finer on males); long, graduated black tail has three narrow white bands. Legs are orange-yellow. Note bright orange cere, lores, and eye-rings and pale brown to pale hazel eyes.

b. Juvenile. Most show dark head and upperparts, whitish underparts with a variable amount of coarser gray barring, and a long, graduated black tail that shows three or four narrow white bands. Most show a whitish hind collar. Cere, lores, and eye-rings are dull yellow-green. Eyes are dark brown, and legs are dull yellow to yellow.

c. Juvenile. Some juveniles have a buffy wash on the hind collar and underparts. This individual is less heavily barred, with barring only on the flanks. Older juveniles can start to show bright face skin. Note the hint of an owl-like facial ring.

d. Adult female. Similar to adult males, but with browner head and upperparts and coarser barring on underparts.

e. Adult flying. White flight feathers show narrow dark banding, and white underwing coverts are finely barred gray. White undertail coverts are finely barred gray; sometimes lacking on males. Blackish tail shows three narrow white bands. Note short, paddle-shaped wings and bright face coloration.

f. Rufous-morph juvenile. Uncommon. Some have rufous heads, upperparts, and wide breast bands and narrowly barred underparts.

2 SLATY-BACKED FOREST-FALCON

Micrastur mirandollei PAGE 262

Somewhat larger than 1 and smaller than 3.

a. Adult flying. Throat and underparts are unmarked white. Flight feathers are heavily barred, more whitish and less heavily barred on bases; shows unmarked or lightly marked white underwing coverts. Blackish tail shows two or three wider pale bands and graduated tip.

b. Juvenile perched. Juveniles have slate-gray head and upperparts, white unmarked throat, and white underparts with scaly brownish mottling, heavier on breast. Black tail has three pale gray bands with a narrow darker band in the center of each. Cere, lores, and eye-rings are bright yellow, bleeding onto the base of the otherwise dark beak, but some have a mostly yellow beak. Eyes are dark brown.

c. Adult. Adults have a slate-gray head and upperparts, a white unmarked throat and underparts, and a black tail with two or three pale gray bands with a narrow darker band in the center of each. Cere, lores, and eye-rings are bright yellow, beak is black, and eyes are dark brown.

3 COLLARED FOREST-FALCON

Micrastur semitorquatus PAGE 264

Largest, with a relatively longer graduated tail. Three color morphs: light (most), buffy (some), and dark (fewer). Light and buffy morphs show a distinctive nuchal collar.

a. Juvenile buffy. Color and pattern are similar to buffy adult, but upperparts are browner with extensive rufous feather edging, underparts are marked with dark brown spotting on breast and barring on flanks, and tail has one or two additional white bands. Legs are dull yellow. Face skin is dull green; cere and beak of fresh juvenile are yellow with smudgy dusky base.

b. Adult flying. Flight feathers are heavily barred, more whitish and less heavily barred on bases; underwing coverts are unmarked white. Blackish tail is graduated and shows four or five narrow white bands.

c. Adult light. Head and upperparts are blackish, and throat, cheeks, hind collar, and underparts are white. Long black tail shows three narrow white bands. Eyes are dark brown, and cere, lores, and eye-rings are dull greenish-yellow. Legs are yellow to orange-yellow.

d. Adult dark. Overall slaty-black; lacks pale cheeks and hind collar. Black undertail coverts show narrow white barring. Tail and flight feathers are like those of light adult.

e. Juvenile dark. Color and pattern are similar to dark adult, but upperparts are browner with some rufous feather edging, belly is narrowly barred rufous, and tail has one or two additional white bands. Legs are dull yellow. Colors of soft parts are like those of other juvenile color morphs. This individual shows an adult-like dark beak.

f. Adult buffy. Similar to light adult, but with buffy replacing white in the plumage. Some adults of any color morph can show a narrow pale superciliary line.

g. Juvenile light. Color and pattern are similar to light adult, but upperparts are browner with some rufous feather edging; underparts are boldly barred dark gray, and tail has one or two additional white bands. Cheek patch and hind collar are less obvious. Legs are dull yellow. Colors of soft parts are like other juvenile color morphs. This individual shows beak color of fresh juvenile.

PLATE 29

APLOMADO FALCON, AMERICAN KESTREL, AND MERLIN

Three small falcons.

1 APLOMADO FALCON

Falco femoralis PAGE 284

Small, colorful, long-winged and long-tailed falcon. Bold face pattern of wide pale superciliary line and narrow dark eye-line and mustache marks, wide dark belly band, and long tail are distinctive. Wingtips fall short of tail tip on perched falcons.

a. First-plumage male. Like adult male, but with several short dark breast streaks.

b. Adult male. Adults have lead-colored napes and upperparts and white barring on belly bands. In fresh plumage, head and breast show rufous wash. Males have unmarked breasts.

c. Adult female. Like male but with dark breast streaks. Head and breast show no rufous in faded plumage. White barring on belly band is often narrower than that of males.

d. Juvenile gliding. Similar to adult, but more buffy, with wider and more extensive breast streaking. Long narrow wings and long tail are distinctive.

e. Juvenile perched. Like adult but dark brown above; breast streaking is wider and more extensive, and belly band has no white barring.

f. Adult gliding. Bold face pattern, narrow white band on rear edge of wings, and long tail are distinctive. Tail shows many (more than four) narrow white bands.

g. Adult male soaring. Patterns of bold face and underparts are distinctive. Dark underwings show narrow white band on rear wings. Note narrow white band on trailing edge of dark underwings.

2 AMERICAN KESTREL

Falco sparverius PAGE 276

Small and colorful falcon. Two black mustache marks are distinctive. All show pale underwings. Wingtips fall short of tail tip on perched kestrels.

a. Adult male soaring. Rufous tail with wide black subterminal band and double mustache marks are distinctive. Note the row of white dots on the tips of the primaries, noticeable on backlit underwings.

b. Adult female hovering. All regularly hover and kite. Head pattern is same as adult male. Female tails are reddish-brown with narrow dark brown bands and wide subterminal band.

c. Adult male perched. Adults have unmarked breast. Males show black spotting on the belly, blue-gray wing coverts, and rufous back barred on the lower half. Note "false eyes" on nape.

d. Juvenile male perched. Similar to adult male, but with dark breast streaking, back barred up to the nape, and paler nape.

e. Adult female perched. Head like males'. Females have reddish-brown upperparts and tail with darker brown barring, and reddish-brown streaking on buffy underparts.

f. Adult female perched, *isabellinus*. Similar to nominate but smaller, and streaking on underparts is faint, hardly noticeable.

g. Adult male perched, *isabellinus*. Similar to nominate but smaller; underparts are more rufous and show little or no spotting, and back shows little or no barring.

3 MERLIN

Falco columbarius PAGE 280

Small, dashing, compact falcon. Mustache marks are faint or absent. Wingtips fall short of tail tip on perched Merlins. Dark tail has three or four pale bands. Adult females and juveniles are usually not separable in the field.

a. Adult female Taiga soaring. Underparts are heavily streaked, and underwings are dark. Dark tail shows three or four pale bands. Note the white throat. Juveniles are alike.

b. Adult female Taiga flying. Upperparts are dark, and underparts are heavily streaked. Adult females have grayish uppertail coverts, those of juveniles are brown.

c. Adult Prairie gliding. Overall paler than Taigas, with more and wider pale tail bands.

d. Adult male Taiga perched. Upperparts are slate-blue, and streaked underparts are rufous-buff, but legs are rufous. Black tail shows several gray bands.

e. Adult male Prairie perched. Overall paler than Taigas, with a hint of a pale collar. Upperparts are gray, and gray tail bands are wider.

f. Adult female Taiga perched. Upperparts are dark brown, and underparts are heavily streaked. Juveniles are alike. Dark brown tail shows three narrow buffy bands.

g. Adult female Prairie perched. Similar to Taiga female, but overall paler, with sandy upperparts, hint of a pale collar, and wider pale tail bands.

PLATE 30

BAT FALCON AND ORANGE-BREASTED FALCON

Two similar dark rain forest falcons: one small and one medium-sized.

1 BAT FALCON

Falco rufigularis PAGE 288

Fairly common small, compact falcon lacking wide rufous band across the breast. Wingtips fall just short of tail tip on perched falcons. Head and feet are relatively smaller than those of 2.

a. Adult female. Adults have dark hoods, white breast that extends below cheek onto the nape, dark underparts with narrow white barring, and unmarked rufous leg feathers and undertail coverts. Note the orangish-yellow eye-ring, cere, and legs of adults.

b. Adult male. Sexes are alike in plumage, but males are noticeably smaller

c. Juvenile female. Similar to adults, but with wider, rufous-buff feather edges on dark underparts, short dark streaks on breast and on rufous belly, wider white tips on tail feathers, and duller rufous, dark-streaked leg feathers and dark-barred undertail coverts (similar to those of 2). Juveniles have yellow eye-rings, ceres, and legs.

d. Adult. All dark below except for white throat and rufous belly. Note the dark hood and unmarked rufous undertail coverts.

e. Adult. All dark above, sometimes showing white on cheeks. Tail bands are usually not visible.

f. Juvenile. Almost like adult, but some rufous-buff feather edges on upperparts are noticeable when seen well in good light.

g. Juvenile. Similar to adults, but with wider, rufous-buff feather edges on dark underparts, short dark streaks on breast and on rufous belly, duller rufous and marked leg feathers, and dark barred undertail coverts.

h. Head-on. Often glides, swift-like, with wings below horizontal.

2 ORANGE-BREASTED FALCON

Falco deiroleucus PAGE 291

Rare and local medium-sized falcon. Distinguished from the much more common 1 by larger size, relatively larger head and feet, wider base of wings, white throat contrasting with rufous upper breast, and darkly barred undertail coverts (juvenile Bat Falcons only). Wingtips fall just short of tail tip on perched falcons.

a. Adult. All dark above, sometimes showing white on cheeks. Narrow white tail bands are often visible. Note the longer head and neck compared to that of 1f.

b. Juvenile. All dark above, sometimes showing white on cheeks. Note bluish eye-ring and cere. Narrow white tail bands are often visible.

c. Juvenile female. Somewhat similar to adults, but with much duller buffy-rufous on breast and vent and brownish band across underparts. Juveniles have bluish to dull yellow eye-rings and ceres and yellow legs. Buffy-rufous undertail coverts are barred black.

d. Adult male. All dark below except for white throat that contrasts with rufous breast band. Other differences from 1 are darkly barred undertail coverts, larger head and neck, and wider base of wings. Legs are sometimes orange, especially on males.

e. Juvenile female. Somewhat similar to adults, but with much duller buffy-rufous on breast and vent, brownish band across underparts, darkly marked leg feathers, and barred undertail coverts. Juveniles have bluish to dull yellow eye-rings and ceres and yellow legs.

f. Adult female. Adults have dark hoods, white throat, and upper breast that contrasts with wide rufous breast band, black underparts with narrow white barring, unmarked rufous leg feathers, and darkly barred undertail coverts. White barring on dark cummerbund is wider than that of adult 1; compare to 1a and 1b. Note the orangish-yellow eye-ring and cere and orange-yellow legs of adults. Females often show faint dark streaks on the rufous breast.

g. Adult male. Sexes are alike in plumage, but males are noticeably smaller, lack any streaking on the breast, and usually have more orangish eye-rings, ceres, and legs.

h. Head-on. Wings are usually held horizontal when flying, sometimes with wingtips held down.

PLATE 31

PRAIRIE FALCON AND PEREGRINE FALCON

Two large falcons: one pale and one dark.

1 PRAIRIE FALCON

Falco mexicanus PAGE 298

Large pale, long-tailed falcon. White areas between eyes and dark ear patches, dark axillars (wing pits), and sparsely marked underparts are distinctive. Head is blocky, with narrow pale superciliary lines and long narrow mustache stripes. Wingtips fall short of tail tip on perched falcons.

a. Head-on. Often glides with wrists lowered and wingtips curved a bit upward.

b. Adult male. All show pale underparts and underwings and dark axillars. Adults have spotted underparts, barred flanks, and yellow-orange legs, and adult males have pale median wing coverts. Males are usually less heavily marked.

c. Juvenile female. Similar to adult from below in flight, but underparts are narrowly streaked, and legs are pale yellow to bluish. Juveniles' median secondary and primary underwing coverts are usually dark.

d. Juvenile female. Cere is blue, and legs are pale yellow to pale bluish. Dark brown upperparts have pale feather edges but no pale cross-barring. Underparts are narrowly streaked.

e. Adult male. Cere and legs are yellow-orange. Sandy upperparts show pale cross-barring and edges, and underparts are spotted, with barred flanks. Note dark patches on sides of upper breast.

f. Adult. Upperparts are a paler, sandy brown, with paler uppertail.

g. Juvenile. Upperparts are dark brown, with paler uppertail.

h. Adult female. Like adult male, but median coverts are dark.

2 PEREGRINE FALCON

Falco peregrinus PAGE 294

Large, dark, long-winged falcon. Dark hooded head, wide dark mustache stripes, and dark underwings are distinctive. Wingtips extend to, or almost to, tail tip on perched Peregrines.

a. Juvenile male *tundrius*. Juveniles have uniformly dark underwings and lightly to heavily streaked underparts. Mustache stripes are narrower on *tundrius*. Dark tail shows narrow buffy bands.

b. Adult male *anatum*. Adults have blue-gray upperparts, white breasts, and barred flanks and belly. *Anatum* adults have blackish hoods and often show a peachy bloom on the breast, heavier on females. Adult cere and eye-rings are yellow to orange-yellow.

c. Adult. Adult upperparts are blue-gray, with paler lower back and tail coverts and darker tail.

d. Juvenile *tundrius*. Juveniles' upperparts are uniformly dark brown, often with narrow pale feather edges. Juvenile *tundrius* have narrow mustache stripes and often pale crowns. Dark tail shows narrow buffy bands and wide white tip.

e. Adult female *anatum*. Adults' underwings are paler than juveniles'. Adults' dark hood, white breast, and barred belly are distinctive. Note whitish undertail with darker band on tip.

f. Adult female *tundrius*. Like adult *anatum*, but head is dark gray, mustache is narrower, white cheek patches are larger, and underparts are less heavily marked. Adult males have less heavily marked underparts than do adult females. Males usually lack peachy bloom on underparts.

g. Juvenile male *anatum*. Hooded head is dark brown, and rufous-buff underparts are heavily streaked. Leg feathers are marked with dark barring or chevrons. Juveniles have bluish to pale yellow (later) eye-rings and ceres.

h. Juvenile female *tundrius*. Similar to *anatum* juveniles, but head usually shows pale superciliary lines, large white cheek patches, and narrower mustache stripes. Underparts are whitish, not rufous-buff. Leg feathers are streaked.

i. Juvenile *tundrius*, pale head. Some have very pale heads, but dark eye-line joins the eye (unlike Prairie Falcon; see 1d and 1e). Streaking on underparts is also sparse.

j. Head-on. Often glides with wrists lowered and wingtips curved a bit upward.

PLATE 32

VAGRANTS AND INTRODUCED

Three vagrants and one introduced.

1 CALIFORNIA CONDOR

Gymnogyps californianus (not to scale) PAGE 97
Huge, mostly blackish vulture, introduced and rare in Baja California, Mexico. All have patagial markers with large number-and-letter combinations written on them, usually on both wings. Note very broad wings and the deeply slotted primary remiges.
a. Juvenile. Head is dark, and underwing coverts are whitish.
b. Adult. Head is orange, and underwing coverts are white. Note white legs.
c. Juvenile. Overall black above, with narrow white band at base of secondaries.
d. Adult. Overall black above, with white band at base of the secondaries and white streaks down the secondaries. Note square tail.
e. Adult. Overall black except for orange head and pinkish neck, white bars on upperwings, and whitish on uppersides of secondaries.
f. Older immature head. Head gradually changes from blackish to orange over four to five years.
g. Juvenile. Overall black except for narrow whitish line at base of upper secondaries.

2 HARLAN'S HAWK

Buteo (jamaicensis) harlani PAGE 231
Large buteo that barely reaches n Mexico in the northern winter. Tails of adults are highly variable. Wingtips of perched hawks fall short of tail tip. Occurs in three color morphs: dark (85%), intermediate (5%), and light (10%).
a. Adult light morph. Uppersides are blackish with noticeable white crown streaks, pied face pattern, and pale gray uppertail that often shows dark mottling and a little rufous near the tip. Note that some show white "headlights" at the base of the forewings.
b. Adult dark morph. Overall blackish with a variable amount of white markings on throat and breast. Tail below is whitish with irregular dark subterminal band. Outer primaries are usually barred.
c. Adult tail variation. Adult tails are highly variable; no two seem exactly the same.
d. Adult dark morph. Overall blackish with a variable amount of white markings on head and upperparts. Some adults show white spots at the base of the forewings.
e. Juvenile intermediate morph. Blackish underparts are heavily and uniformly mottled white. Throat is whitish or with white streaking. Outer primaries are usually barred. Tail bands are even or wavy or chevron-shaped, usually with the last band showing a spike or hourglass toward the tip.
f. Juvenile dark morph. Blackish upperparts are heavily mottled white. Breast is more heavily marked compared to belly. Throat is whitish or with white streaking. Tail bands are even or wavy or chevron-shaped, usually with the last band showing a spike or "hourglass" at the tip.
g. Adult dark morph. Overall blackish with a variable amount of white markings on face, crown, and breast. Tail above is grayish with dark mottling and an irregular wide dark subterminal band.
h. Adult light morph. Overall black and white, with strong face pattern and streaked crown. Lacks any buffy tones on underparts. Undertail is whitish with dark subterminal band.

3 LONG-WINGED HARRIER

Circus buffoni PAGE 144
Vagrant from South America. Two records of light-morph harriers in eastern Panama. All show pale facial ring and white uppertail coverts. Only the light morph is depicted.
a. Juvenile flying. Lightly marked individual. Head and upperparts are brown, and pale buffy underparts show faint narrow streaking on upper breast and flanks. Primaries are gray, and secondaries are brown above, both with narrow dark banding. Remiges below are pale gray and show narrow dark banding. Wing linings are heavily marked. Long tail shows blackish and pale gray (or brown) bands of equal width above and buffy and dark brown bands of equal width below. Eyes are dark brown.
b. Adult male flying. Head and upperparts are black, with noticeable facial disk, and white underparts are unmarked. White wing linings are narrowly barred black. Remiges and tail are banded black and gray. Eyes are yellow.
c. Adult female. Dark gray head shows bold white face pattern. Dark gray upperparts contrast with black-banded silvery-gray remiges and tail. White underparts are sparingly marked, with some streaking across the chest. Eyes are dark brown.
d. Juvenile. Heavily marked individual. Some juveniles are heavily streaked. Eyes are dark brown.

4 GRAY-BELLIED HAWK

Accipiter poliogaster PAGE 146
Vagrant from South America. A juvenile was recorded in Costa Rica.
a. Juvenile. Plumage mimics adult Ornate Hawk-Eagle. See plate 25, figs. 3d and 3h. Tarsi are unfeathered.
b. Juvenile. Long legs are unfeathered. Lacks long crest. Much smaller than Ornate Hawk-Eagle.
c. Adult. Head is blackish, and cheeks can be black, as shown, or gray. Upperparts are dark gray, and underparts are white. Black tail shows three narrow gray bands. Eyes are dark. Cere, gape, and face skin are orange-yellow.

BLACK VULTURE PLATE 2
Coragyps atratus

ZOPILOTE NEGRO, GALLINAZO

LENGTH 59–74 cm   **WINGSPREAD** 141–160 cm   **WEIGHT** 1.7–2.3 kg

IDENTIFICATION SUMMARY Distinctive large, jet black vultures. Short black tail, white primary patches, and whitish legs are diagnostic. Adult and juvenile plumages are similar. Wingtips reach tail tip, and primaries extend barely beyond secondaries (short primary projection) on perched vultures.

TAXONOMY AND GEOGRAPHIC VARIATION *C. a. atratus* occurs in n Mexico, and *C. a. brasiliensis* in s Mexico and Central America. Alleged racial differences are not detectable in the field.

SIMILAR SPECIES (1) **Turkey Vulture** (plate 1) is brownish-black, not jet black, on the back; has longer tail and silvery flight feathers on underwings; and flies with slower wing beats and with wings held in a stronger dihedral. On perched Turkey Vultures, primaries extend noticeably beyond secondaries. Adult Turkey Vultures have red heads (and white nape in Panama and Costa Rica).

STATUS AND DISTRIBUTION Common throughout except for Baja California, high mountain chain in c Mexico, and nc Mexico.

HABITAT Occurs in a variety of open habitats, but only edges of forests.

BEHAVIOR Black Vultures are scavengers that subsist on large carcasses. They are reported to catch fish; to attack live prey, especially newborn pigs and other livestock and even skunks; and to eat oil palm fruit. They locate prey not by smell, but by watching other scavengers and by frequenting abundant food sources, such as dumps and slaughterhouses. They are more gregarious and aggressive than Turkey Vultures and dominant over them at carcasses, mainly because of their greater numbers.

They soar with wings held in a slight dihedral and glide with wings level to somewhat elevated. Their active flight is distinctive: three to five shallow, rapid, and stiff wing beats with wings thrust forward, followed by a short period of glide. Their wing loading is heavier than that of Turkey Vultures, and they require stronger thermals for soaring and therefore do not begin flying in the morning until an hour or so after Turkey Vultures. Black Vultures usually soar higher than Turkey Vultures and, like that species, often bow their wings downward in a flex until the tips almost meet. They regularly soar and glide with legs dangling.

Black Vultures form communal night roosts, often with Turkey Vultures. Breeding is solitary, with nesting recorded on the ground under thick vegetation and a tree root; under a rock overhang; and in

a small cave, an abandoned stick nest of other birds, and even an abandoned building.

Vocalizations reported are restricted to hisses and grunts. Feathered nape can be extended onto crown in colder weather.

MOLT Annual molt is complete, from spring to autumn. Second prebasic molt begins when about one year old and is completed before winter.

DESCRIPTION Sexes are alike in plumage and size. Eyes are dark brown. Juveniles differ from adults only by head skin beak color. **Adult.** Gray head skin is featherless and wrinkled, becoming more wrinkled with age. Body and wing coverts are jet black, and back and upperwing coverts show an iridescence under some light conditions. Black upperwings and underwings show large pale gray primary patches, larger on underwings. Beak is dusky with yellowish tip. **Juvenile.** Very similar to adult, but head and neck skin are black and smooth, not wrinkled; beak is completely dusky; and body plumage is somewhat less iridescent than that of adult. Feathered nape extends farther up hind neck than does that of adult.

FINE POINTS The down of Black Vulture chicks is buffy; that of Turkey Vulture chicks is white.

UNUSUAL PLUMAGES Published sight records are of one fawn-colored (amelanistic) and another mostly white (partial-albino) individual. An adult from c Mexico had a reddish head, which could be a color anomaly caused by DNA or disease or a hybrid with Turkey Vulture.

Black Vulture. Adult. Adults have wrinkled heads and dark beaks. Note the whitish legs. Brazil. August 2012. WSC

BELOW: Black Vulture. Adult. All show pale gray primary panels and white legs. Adults have gray wrinkled heads. Connecticut USA. December 2003. JIM ZIPP

HYBRIDS The hybrid reported in 1937 in *The Auk* between this species and a Turkey Vulture from Louisiana was a practical joke, apparent when the red paint wore off the head on a specimen of a normal Black Vulture. See Unusual Plumages above.

ETYMOLOGY "Vulture" comes from the Latin *vulturus*, "tearer," a reference to its manner of eating carrion. *Coragyps* comes from the Greek *korax*, "raven," and *gyps*, "vulture." The Latin *atratus* means "clothed in black, as for mourning."

Black Vulture. Juvenile. Juveniles have smooth blackish heads covered in fine dark down. Florida USA. January 1995. wsc

TURKEY VULTURE PLATE 1

Cathartes aura

ZOPILOTE CABEZA ROJA

LENGTH 62–72 cm **WINGSPREAD** 160–181 cm **WEIGHT** 1.6–2.4 kg

IDENTIFICATION SUMMARY Large, brownish-black vultures. Adults have red heads. Underwings are two-toned: silvery flight feathers contrast with blackish coverts. They fly with wings in a strong dihedral. Wingtips reach tail tip on perched vultures.

TAXONOMY AND GEOGRAPHIC VARIATION Three subspecies occur in North America, Mexico, and n Central America (*C. a. septentrionalis*, *C. a. meridionalis*, and *C. a. aura*, respectively), but they are almost alike. Breeding vultures in Panama and Costa Rica are *C. a. ruficollis*, the n South American form, and are distinguished by their white napes and somewhat smaller size.

SIMILAR SPECIES (1) **Lesser Yellow-headed Vulture** (plate 1) perched shows wingtips that extend far beyond the tail tip. In flight, has proportionally longer wings, adults have bluish and yellow on the head, and juvenile has paler gray head, compared to Turkey Vultures.

(2) **Zone-tailed Hawk** (plate 22) is an excellent Turkey Vulture mimic; see under that species for distinctions.

(3) **Black Vulture** (plate 2) is somewhat similar; see under that species for distinctions.

(4) **White-tailed Hawk** (plate 20) juvenile is somewhat similar; see under that species for distinctions.

STATUS AND DISTRIBUTION Fairly common to common throughout.

BELOW LEFT: Turkey Vulture. Adult. Wingtips reach tail tip and legs are pinkish on all. California USA. July 1994. WSC

BELOW RIGHT: Turkey Vulture. Adult. Underwings are two-toned on all. Texas USA. April 2006. WSC

ABOVE: Turkey Vulture. Adult. They often sun with wings spread. Florida USA. January 1995. WSC

Turkey Vulture. Juvenile. Juveniles lack red heads. Connecticut USA. October 2012.
JIM ZIPP

HABITAT Occurs in almost all habitats.

BEHAVIOR Turkey Vultures spend much of the day soaring at moderate heights, searching for carrion. They are able to locate carrion by smell as well as by sight. While carrion is the vulture's usual fare, there are isolated reports of Turkey Vultures catching live fish and attacking live animals, usually ones that are sick or in some way incapacitated. Studies have shown that Turkey Vultures eat smaller prey than Black Vultures, but Blacks dominate at carcasses because of their numbers.

Active flight is with slow, deep, deliberate wing beats on flexible wings. They soar and glide with wings held in a strong dihedral, often rocking or teetering from side to side. When they are gliding in strong winds, the wings are pulled back with only a slight dihedral.

Turkey Vultures often bow their wings downward in a flex until the tips almost meet.

Turkey Vultures do not construct a nest. They lay eggs in small caves, on the ground under heavy vegetation, in large tree cavities, and even in abandoned buildings.

Communal night roosts of up to several hundred birds, sometimes including other vultures, form on buildings, on radio towers, and in large trees. Vocalizations reported are limited to grunts and hisses.

MOLT Annual molt of adults is usually complete, ending with the inner three primaries. Second prebasic molt begins when vultures are about a year old and is completed with P1–P3 replaced twice. Molt sequence of flight and tail feathers is almost like that of accipitrid raptors, except for ending the primary molt with P3.

DESCRIPTION Sexes are alike in plumage and size. Adult plumage is attained in one year, but first-plumage adults show dark tip on beak. Legs are pinkish to pale reddish, but are often whitened by defecation. Eyes are gray-brown. **Adult.** Featherless head is red and covered with a variable number of whitish warts on the lores; reddish neck is wrinkled (nape is whitish on residents in Costa Rica and Panama). Entire body is brownish-black, with upperwing coverts and lower breast and belly somewhat browner. Neck and back have a purplish iridescence. From below, flight feathers are silvery, and wing coverts are black, forming two-toned underwing pattern. Underside of long tail is silvery, but darker than flight feathers. Beak is ivory. **Juvenile.** Similar to adult, but head is dusky, lacks wrinkles and warts, and is covered with blackish fuzzy down on crown and hind neck. Face gradually turns pink with age and is usually red when vultures are almost a year old. Beak is dusky, with a pale base that gradually enlarges toward tip. Neck and back lack iridescence. Upperwing coverts have wide, buffy feather edges. Silvery undersides of flight feathers are a bit darker that those of adults, but underwings still show a two-toned pattern. **Basic II (second plumage).** Almost like adult, but with few, if any, warts on head and dark tip to beak.

FINE POINTS Nestling's down is whitish; Black Vulture nestling's down is buffy.

UNUSUAL PLUMAGES Amelanistic vultures are light gray. Partial albinos with some to mostly white feathers, including some with only the outer primaries white, have also been reported.

HYBRIDS Hybrid with Black Vulture from Louisiana reported in *The Auk* was a hoax.

ETYMOLOGY *Cathartes* is from the Greek *kathartes*, "purifier." The species name, *aura*, possibly derives from Latin *aurum*, "gold," because of the head color of museum specimens but it could be from a Latinized version of a native South American word for "vulture."

LESSER YELLOW-HEADED VULTURE PLATE 1

Cathartes burrovianus

AURA CABEZA AMARILLA

LENGTH 53–65 cm **WINGSPREAD** 150–165 cm **WEIGHT** 0.9–1.6 kg

IDENTIFICATION SUMMARY Large, slender, black vultures. They show some yellow on the head. Juvenile plumage is similar to adults. Underwings are two-toned: silvery flight feathers contrast with blackish coverts. Wingtips extend far beyond tail tip on perched vultures, and bright pale shaft streaks are visible on longest primaries on upperwings of flying vultures.

TAXONOMY AND GEOGRAPHIC VARIATION Monotypic, with no geographic variation.

SIMILAR SPECIES (1) **Turkey Vulture** (plate 1) is somewhat similar; see under that species for distinctions.

(2) **Zone-tailed Hawk** (plate 22) is a mimic of Turkey Vultures and therefore similar to this species; see under that species for distinctions.

STATUS AND DISTRIBUTION Fairly common, but local in the wet lowlands on the Caribbean slope of c Mexico to c Nicaragua, including the Yucatán Peninsula; on the Pacific slope from Oaxaca to n Costa Rica; and on the Pacific slope from sw Costa Rica into ec Panama.

HABITAT Occurs in wetter open areas, including marshes, wet pastures, savannahs, and mangroves.

Lesser Yellow-headed Vulture. Adult. Adults have colorful heads. Wingtips of all extend beyond tail tip. Belize. December 2012. WSC

BEHAVIOR They forage for carrion by gliding low over wet open areas and are able to locate carrion by smell as well as by sight. They also spend lots of time perched on the ground or on a low fence post.

Active flight is with slow, deep, deliberate wing beats on flexible wings. They soar and glide with wings in a strong dihedral, often rocking or teetering from side to side, rarely soaring high. Compared to Turkey Vultures, they fly faster and more aggressively and usually closer to the ground. When the bird is gliding in strong winds, the wings are pulled back with only a slight dihedral. Adults' head color varies with mood; the head is redder when the vulture is excited. Cathartid vultures often bow their wings downward in a flex until the tips almost meet.

Pair lays eggs on the ground in dense vegetation or in tree hollows.

Communal night roosts in trees consist of small groups. Vocalizations reported are limited to grunts and hisses.

MOLT Annual molt of adults is usually complete. Second prebasic molt begins when about a year old and is completed before winter; the inner three primaries are replaced twice. Molt sequence of flight and tail feathers is almost like that of accipitrid raptors.

DESCRIPTION Sexes are alike in plumage and size. All show bright pale shaft streaks on outer six primaries, which are paler than

Lesser Yellow-headed Vulture. Adult. Adults have colorful heads. Underwings are two-toned. Mexico. October 2007. wsc

Lesser Yellow-headed Vulture. Juvenile. Juveniles have pale gray heads. Wingtips of all extend beyond tail tip. French Guinea. May 2011. MICHEL GIRAUD-AUDINE

secondaries. Legs are pinkish to pale reddish, but are often whitened by defecation. **Adult.** Featherless head consists of bluish crown and lores, yellow face, and red nape and neck. Body and coverts are black. From below, flight feathers are silvery, and wing coverts are black, forming two-toned underwing pattern. Underside of long tail is silvery, but darker than flight feathers. Eyes are gray-brown. Beak is ivory. **Juvenile.** Similar to adult, but head is pale gray and is covered with whitish fuzzy down on crown and hind neck. Face gradually colors with age. Beak is dusky, with a pale base that gradually enlarges with age toward tip. Upperwing coverts have buffy feather edges. Eyes are brownish-gray.

FINE POINTS Head color changes according to mood, from pale yellow to nearly as red as that of a Turkey Vulture.

UNUSUAL PLUMAGES No unusual plumages have been reported.

HYBRIDS None reported.

ETYMOLOGY *Cathartes* is from the Greek *kathartes*, "purifier." The species name, *burrovianus*, is after little-known Dr. M. Burrough. "Yellow-headed" is obviously for the head color. "Lesser" differentiates this species from the larger Greater Yellow-headed Vulture (*G. melambrotus*) of Amazonia. "Vulture" comes from the Latin *vulturus*, "tearer," a reference to its manner of eating carrion.

CALIFORNIA CONDOR PLATE 32
Gymnogyps californianus

CÓNDOR CALIFORNIA

LENGTH 109–127 cm **WINGSPAN** 249–300 cm **WEIGHT** 8.2–14.1 kg

IDENTIFICATION SUMMARY Largest North American raptor, much larger than any eagle. They have been introduced to areas in Baja California, Mexico. They are overall black, except for bold white triangular patches on the underwings and a narrow white bar on the uppersides of their long, broad wings. Wingtips show very long "fingers." The relatively short tail has a square tip.

TAXONOMY AND GEOGRAPHIC VARIATION Monotypic, with no geographic variation.

SIMILAR SPECIES No other raptors are similar.

STATUS AND DISTRIBUTION Recently reintroduced back into the wild in Sierra de San Pedro Mártir National Park in Baja California, Mexico, with a 2014 population of 29 individuals. All have number-letter patagial markers.

HABITAT Occurs in Baja California in pine-covered mountains.

BEHAVIOR These huge vultures subsist entirely on carrion. They prefer large carcasses, such as deer and cattle. They usually leave night roosts to begin foraging late in the morning after strong thermals form, often returning to a known carcass. Condors spend much time perched and, when thermals are available, begin soaring both singly and in small groups. Sunning behavior, common among vultures, has been noted. Perched birds spread their wings out and face away from the sun. This species is usually dominant over other scavengers at carcasses. After feeding, they have a curious habit of rubbing their heads on the ground. The adult's head color varies, depending on the condor's mood.

California Condor. Adult. Adults have puffy orange heads. All have patagial tags. Arizona USA. March 2016. WSC

Active flight is with slow, stiff wing beats. A California Condor soars and glides very steadily, like an airplane, with wings held slightly upraised. When beginning a glide after gaining altitude in a soar, condors take one or two powerful wing beats, bending their wings below the body until the wingtips almost touch.

MOLT Molt is not completed annually, and most remiges are usually replaced every two years. Molt is suspended in winter. Juveniles begin their second prebasic molt in their first spring. Adult plumage is attained after six or seven molts.

DESCRIPTION A huge, large-headed vulture. **Adult.** Reddish-orange to yellow head shows an unusual tuft of short black feathers between and below the eyes. Head and neck often appear puffed out or inflated. Base of neck is reddish in front and pinkish-gray on sides. Black feathers at base of neck form a ruff. Body and upperwing

California Condor. Older immature. Condors in the first three plumages have dark heads. Note molt in the remiges. All have patagial tags. Arizona USA. March 2016. wsc

coverts are sooty black. A small vertical area of red skin in center of breast is usually visible on perched and flying birds. Perched birds show a narrow white bar on folded wings. Flight feathers are black, but newer secondaries have a silvery wash on upper surface, contrasting with darker old feathers. White underwing coverts form a large narrow triangular patch on underwing. Legs are whitish. Black tail has square tip. Eyes are scarlet, and beak is horn-colored. **Juvenile.** Similar to adult, but narrower dusky head and neck are covered with dusky, fuzzy down and lack the puffy appearance of adults. Upperwings show a narrow dull whitish bar. Whitish area on underwing coverts is heavily mottled with black. Beak is black, and eyes are gray-brown. **Older immatures.** Condors mature over a period of seven or more years. Birds gradually change from juvenile to adult plumage beginning when they are around two years of age. **Note.** All condors in the wild have patagial markers.

FINE POINTS The outer six primaries of soaring condors are very long and narrow, and in flight are bent upward and appear more brush-like and noticeable than those of other large soaring raptors. This species takes from 13 to 17 seconds to complete a circle when soaring, longer than other raptors. Tails of flying birds sometimes appear to have wedge-shaped tips.

UNUSUAL PLUMAGES No unusual plumages have been described.

HYBRIDS No hybrids have been reported.

ETYMOLOGY *Gymnogyps* comes from the Greek *gymnos*, "naked," a reference to the featherless head, and *gyps*, "vulture." The species name is from California.

KING VULTURE PLATE 2
Sarcoramphus papa
ZOPILOTE REY

LENGTH 68–80 cm **WINGSPREAD** 170–200 cm **WEIGHT** 3–3.8 kg

IDENTIFICATION SUMMARY Large black and white vultures. Adults are distinctive, but juvenile is overall dark, becoming progressively whiter with age. Wings are long and broad; tail appears short. Triangular wings are widest at body and taper to tips. Wingtips reach the tail tip, and primaries extend a bit beyond secondaries (short primary projection) on perched vultures.

TAXONOMY AND GEOGRAPHIC VARIATION *Sarcoramphus papa* is monotypic, with no geographic variation.

SIMILAR SPECIES No other raptor is similar to adults. Dark juveniles are somewhat similar in shape to Black Vultures, but have proportionally longer wings and are much larger.

King Vulture. Adult. Adults have colorful heads. Note the full crop. Belize. March 2011. RYAN PHILLIPS-BRRI

King Vulture. Adult. Note triangular shape of black and white wings. Belize. June 2013.
RYAN PHILLIPS-BRRI

STATUS AND DISTRIBUTION Uncommon over forested areas from s and e Mexico to Panama, but absent from highlands of se Mexico to e Honduras and n and c Pacific coasts of Nicaragua.

HABITAT Occurs over tropical forests, occasionally over nearby second growth and open areas.

BEHAVIOR They are usually seen singly, in pairs, or occasionally three or four together, soaring and gliding high above forest, searching for carrion. They occasionally fly over open areas, but usually not far from forest. They're seen occasionally perched inside of forest or at carcasses, where they are dominant over the smaller vultures because of their greater size. They locate prey not by smell, but by watching other scavengers.

Active flight is with slow, deliberate, deep wing beats of stiff wings. They soar with wings held in a slight dihedral and glide with wings level to somewhat elevated or cupped. Their wing loading is heavier than that of smaller vultures, and so they require stronger thermals or winds for soaring. They usually soar higher than the smaller vultures and, like them, often bow their wings downward in a flex until the tips almost meet.

Breeding is solitary. Pairs nest in forests, where they lay one or two eggs in a scrape in leaf litter in dense vegetation, beside a tree trunk, or in a hollow log or stump. Vocalizations reported are restricted to hisses and grunts.

MOLT Annual molt isn't complete; especially, not all remiges and rectrices are replaced. Second prebasic molt begins when vultures are about one year old.

DESCRIPTION Sexes are alike in plumage and size. Legs are whitish from defecation. Adult plumage is attained after five years. **Adult.** Overall black and white; colorful magenta head has red crown patch, long orangish wattles that hang down over the orangish nares, red orbital rings, white eyes, an orangish and dark beak, paler gray

King Vulture. Juvenile. Juveniles have dark heads. Belize. March 2007. RYAN PHILLIPS-BRRI

wrinkled cheeks, and orange-red and yellow neck. Feathers of lower neck and upper body (ruff) are gray. Feathers of body, legs, and coverts are white, often with a pinkish bloom on the upperparts, and flight and tail feathers and primary upperwing and uppertail coverts are black with a silvery cast. Outer edge of each secondary shows whitish line on the uppersides. **Juvenile.** Overall dark sooty-gray. Head shows black beak, often with orange smudges, blackish cere, and short, erect wattle. Smooth head is covered with patches of short black down, and neck is blackish. Wing linings are somewhat paler than remiges. Some show white areas on inner legs and undertail coverts, and white streaking on axillars. Eyes are dark brown. **Basic II (second plumage).** Appears much like juvenile, but beak is now dull reddish, small erect wattle and cere are blackish, wing linings show narrow white streaking, axillars are white, and new secondaries and rectrices are black and broader than retained juvenile ones, which are now brown. Eyes are pale gray-brown. **Basic III.** Appears much like the previous two plumages, but beak is reddish, and longer erect wattle and cere are now dusky and yellowish, and areas of the neck are corrugated. Some have mostly white belly, undertail coverts, and leg feathers. New secondaries have a silvery cast on their uppersides. Ruff is blackish. Eyes are pale. **Basic IV.** Appears somewhat different from previous plumages in that the head and neck are becoming adult-like, and underparts are completely white. Ruff is blackish, upperwing coverts are dark, and wattle is longer and erect. But white upperwing coverts show some black speckling, and longer yellow wattle is still erect.

FINE POINTS Full crop is visible as a blackish to pinkish bulb in the center of the breast.

UNUSUAL PLUMAGES No unusual plumages have been described.

HYBRIDS None reported.

ETYMOLOGY "Vulture" comes from the Latin *vulturus*, "tearer," a reference to its manner of eating carrion. *Sarcoramphus* comes from the Greek *sarx*, "flesh," and *ramphos*, "bill," for the fleshy wattle over the beak. The Latin *papa* means "bishop."

King Vulture. Older immature. Upperwing coverts are whitish with black markings and head is becoming adult-like. Belize. March 2007.
RYAN PHILLIPS-BRRI

OSPREY <small>PLATE 3</small>

Pandion haliaetus

AGUILA PESCADORA

LENGTH 53–66 cm   **WINGSPREAD** 149–171 cm   **WEIGHT** 1.0–1.8 kg

IDENTIFICATION SUMMARY Large, long-winged, whitish, eagle-sized raptors, usually found near water. In flight, their gull-like crooked wings and white head with wide black eye-stripe are distinctive. Perched birds appear long-legged and often show a narrow white stripe between shoulders and body; their wingtips extend just beyond tail tip. Tail on distant flying birds can appear orangish.

TAXONOMY AND GEOGRAPHIC VARIATION *P. h. carolinensis* is resident in nw Mexico and a migrant and winter visitor along both coasts. Juveniles and some older immatures remain for their first summer. *P. h. ridgwayi* is a breeding resident along the e Yucatán coast of Mexico and Belize and is smaller and more white-headed.

Osprey. Adult male. Males lack dark necklaces. Wingtips of all extend beyond the tail tip. Texas USA. April 2006. wsc

Osprey. Adult male.
Note the arched wings.
Texas USA. April 2006.
WSC

SIMILAR SPECIES (1) **Bald Eagles** (plate 3) are larger, usually have dark bodies and broad wings, fly with wings flat, and lack black carpal patches on underwings (see Basic III Bald Eagle with eye-stripe).

(2) **Large gulls** can appear very Osprey-like, but are smaller and have shorter, pointed wings, longer necks, and an unbanded tail, and they lack the black carpal patches on the underwing.

(3) **Snail Kite** (plate 7) faded immatures, when perched, show a similar head pattern as perched Ospreys, but they are much smaller, and have streaked underparts.

STATUS AND DISTRIBUTION Fairly common breeding residents on Baja California and the Gulf of California in nw Mexico and along the Caribbean coast of the Yucatán Peninsula and Belize. They are winter visitors throughout except for mountainous areas. Occasionally Ospreys are found at sea far from land.

HABITAT Occurs almost always near water, except when migrating overland. They will fish in almost any body of water: seas, lakes, ponds, rivers, streams, canals, etc.

BEHAVIOR Ospreys are superb fishermen, catching prey with their feet after a spectacular feet-first dive, most often from a hover but sometimes from a glide, usually entering the water completely. They are able to take off from the surface and, after becoming airborne, shake vigorously to remove water. Ospreys always carry fish head

103

forward. Although they are almost exclusively fish-eaters, they also prey on birds, turtles, and small mammals.

Active flight is with slow, steady, shallow wing beats on somewhat flexible wings. They soar and glide with wings crooked in a gull-winged shape, with wrists cocked forward and held above body level and wingtips pointed down and back. But they occasionally soar with wings level. They hover frequently while hunting.

Large stick nests are built in a variety of trees, including large cacti, as well as on buoys and other man-made structures, and even on the ground.

The high, clear, whistled alarm call, repeated continuously, is distinctive.

MOLT They molt during the summer and winter but suspend it during migration; adults also suspend molt while breeding, especially males. Flight feather molt is incomplete, typical of large raptors. Secondaries are usually replaced every other year; juvenile and later secondaries are the same length. Juveniles begin molt on the winter grounds at an age of five to seven months.

DESCRIPTION Sexes are almost alike in plumage, but females are somewhat larger. Cere and legs are dull blue-gray. **Adult.** White head has dark brown central crown and nape patches and wide dark brown eye-lines that extend down the sides of the neck to the shoulders. Caribbean Ospreys (*P. h. ridgwayi*) have white heads with narrow, incomplete eye-lines and lack dark patches. Eyes are bright yellow. Back and upperwing and uppertail coverts are dark brown. Underparts are white, usually with a more or less distinct band of short dark streaks forming a necklace across upper breast; females usually have more streaks, but there is an overlap in this character as some males show a faint necklace, and a few females can be completely clear-breasted. Underwings show light to dark gray flight feathers, with secondaries darker than primaries, black carpal patches, and white underwing coverts with darker greater coverts forming blackish lines. Short tail appears dark with pale bands from above, but pale with dark bands from below. **Juvenile.** Similar to adult, but upperparts appear scaly due to pale tips on most upperwing coverts and back feathers. Head is like that of adult except that it has narrow dark streaking in white areas of crown and nape. Iris is initially red and gradually fades to orange. Recently fledged birds show a rufous wash on nape and breast, but this fades quickly, and these areas are white by autumn migration. Flight feathers have wide pale tips. Undersides of secondaries are paler than those of adults; greater secondary underwing coverts have white tips, so black lines on underwings are not as distinctive as those of adults. Tail has a wider white terminal band than that of adults.

FINE POINTS Ospreys' lack of a supra-orbital ridge makes them look pigeon-headed. In mated pairs, females almost always have the more distinct breast band, but there is overlap in this character as some males can show a faint necklace. Clear-breasted, unmarked adult birds are usually, but not always, males.

UNUSUAL PLUMAGES Birds with a few white feathers in place of normally dark ones have been reported. Melanistic individuals have been reported from Florida and France.

HYBRIDS No hybrids have been reported.

ETYMOLOGY *Pandion* was the name of two mythical kings of Athens; *haliaetus* is from the Greek *hals* and *aetos*, "sea" and "eagle." "Osprey" came from the Latin *ossifragus*, "bone breaker," but this name probably referred originally to another species.

Osprey. Juvenile. Juveniles have whiter secondaries and often show buffy on underwing coverts. Texas USA. September 2007. WSC

GRAY-HEADED KITE PLATE 4

Leptodon cayanensis

MILANO CABEZA GRIS

LENGTH 43–55 cm   **WINGSPAN** 90–110 cm   **WEIGHT** 380–640 g

IDENTIFICATION SUMMARY Fairly large kites that have a distinctive adult plumage. Juvenile plumage is variable, with four different forms. All show slender bill, short legs, small feet, and long tail. They soar on paddle-shaped wings. Primaries project just over halfway down the tail on perched kites.

TAXONOMY AND GEOGRAPHIC VARIATION The nominate subspecies occurs in our area, with no geographic variation.

SIMILAR SPECIES Adults are unlike any other raptor in our area.

(1) **Black and White Eagle** (plate 25) is similar to white-morph juvenile kites in having white head with black crown, white underparts, and dark upperparts. But kites are smaller and have darker eyes, bare tarsi, and yellow lores.

(2) **Collared Forest-Falcon** (plate 28) white-morph adults are similar to whitest juvenile kites, but are smaller and more robust, with shorter tail and bright orange cere and lores; they lack white hind collar.

(3) **Hook-billed Kite** (plate 4) juvenile males are similar to whitest juvenile kites; see under that species for distinctions.

(4) **Ornate Hawk-eagle** (plate 25) juveniles are similar to white-morph juvenile kites in having white underparts and dark upperparts.

BELOW LEFT: Gray-headed Kite. Adult. Note lack of a supra-orbital ridge. French Guinea. March 2013.
MICHEL GIRAUD-AUDINE

BELOW RIGHT: Gray-headed Kite. Adult. Paddle-shaped wings show below heavily banded primaries and dark wing linings. Belize. March 2007.
RYAN PHILLIPS—BRRI

But kites are smaller, have bright orange cere and lores, and lack long white crest and dark barring on leg feathers.

STATUS AND DISTRIBUTION Uncommon and local in lowland to foothill forests from s Mexico to Panama; they avoid highlands.

HABITAT Occurs in a variety of well-watered lowland and foothill forests, including gallery forest and forest edge of water bodies. Occasionally occurs in adjacent semi-open areas.

Gray-headed Kite. Juvenile. White morph. Other juvenile morphs show streaked underparts. Costa Rica. January 2006.
GLENN BARTLEY

BEHAVIOR They feed on a variety of small prey items, which they hunt from inconspicuous perches within the forest. They take many insects, especially cicadas and larvae of wasps and bees, but also eggs and nestling birds, lizards, and frogs. They are particularly unobtrusive, but sometimes perch, especially in the early morning, on a high bare branch.

They soar regularly with wings held level, but they glide on somewhat cupped wings: wrists up and tips down. Powered flight is with slow, soft, deep wing beats, but they can be quite agile and quick in pursuing insects. Display flights start with a glide, and then continue with an abrupt climb with deep, rapid, floppy wing beats, then a glide down with wings held up in a strong dihedral.

They are usually silent except when breeding. The usual call then is a series of trogon-like barks, as well as other repeated calls and a cat-like call. They build a small stick nest high in the forest canopy.

MOLT Not studied. Most likely annual molts are complete. First-plumage adults may show white areas in black underwing coverts.

DESCRIPTION Sexes are alike in plumage. All have slender bill, short legs, small feet, and long tail. **Adult.** Adults have gray cere, lores, eye-ring, and legs. Most have dark eyes, but some show gray eyes. Head is pale gray, upperparts are dark gray, and underparts are white. Underwings show bold wide black and white bands on primaries, indistinct narrow bands on pale secondaries, and black coverts. Long black tail has two bands, grayish above and white below. **Juveniles.** Juveniles have yellow to orange-yellow cere, lores, eye-ring, and legs. Their eyes are yellow to amber. Four different juvenile plumages can be recognized. All show similar tails that have dark brown and pale brown wide bands. Eyes are brown. Cere and legs are yellow to orange. Underwing coverts are marked like underparts. **White morph.** Head is white with dark crown patch. Upperparts are brown with wide white collar. Whitish underparts are unstreaked. **Intermediate light morph.** Head is darker, but throat is white with narrow dark mesial stripe. White collar is narrower, and underparts show narrow dark streaking. **Intermediate dark morph.** Head is dark, but throat is white with wide dark mesial stripe. Collar on hind neck is rufous, and underparts show wide dark streaking. **Dark morph.** Head, neck, throat, and breast are dark, with only hints of rufous collar. The rest of the underparts are heavily streaked.

FINE POINTS Most adults have uniformly black underwing coverts, but some (presumably first-plumage adults or females) show white areas there.

UNUSUAL PLUMAGES No unusual plumages have been reported.

HYBRIDS No hybrids have been reported.

ETYMOLOGY *Leptodon* comes from the Greek *leptos* and *odon*, "thin" and "tooth," in reference to its sharp beak. The species name, *cayanensis*, is after Cayanne, or French Guiana, where the first specimen was collected.

HOOK-BILLED KITE PLATE 4

Chondrohierax uncinatus

GAVILÁN PINTADO

LENGTH 41–51 cm **WINGSPAN** 84–98 cm **WEIGHT** 215–378 g

IDENTIFICATION SUMMARY Large hooked beak and paddle-shaped wings that pinch in to the body on trailing edges. Head shows distinctive greenish lores and bare yellow-orange spots above the eyes. Sexes are similar in size, but have different adult plumages. Juvenile plumage is somewhat similar to that of adult female. Dark-morph kites also have distinct plumages for adult males and females and juveniles. Legs are yellow to yellow-orange. Wingtips extend halfway down the long tail on perched kites.

TAXONOMY AND GEOGRAPHIC VARIATION Widespread race *C. u. uncinatus* occurs throughout. Alleged subspecies *acquilonis* in Mexico hardly differs.

SIMILAR SPECIES (1) **Gray Hawk** (plate 17) adult is more finely barred below than adult male Hook-bill and has dark eyes, more pointed wingtips, and a smaller beak, and its underwings are whitish and lack boldly barred primaries.

(2) **Roadside Hawk** (plate 12) also has pale eyes, paddle-shaped wings, and barred underparts, but lacks large hooked beak and collar on hind neck (adult female) and has uniformly dark or streaked breast. Roadside Hawks in n Mexico usually lack the rufous in primaries of adult female Hook-bills.

BELOW LEFT: Hook-billed Kite. Adult Male. Adult males are overall gray. All show paddle-shaped wings. Texas USA. June 2010. WSC

BELOW RIGHT: Hook-billed Kite. Adult female. Adult females are brown above and marked rufous below. Note the rufous hind collar. All lack supra-orbital ridges. Texas USA. June 2010. WSC

(3) **Snail Kite** (plate 7) adult male is also overall dark gray like dark-morph kites; see under that species for distinctions.

(4) **Plumbeous Hawk** (plate 12) is also overall dark gray like dark-morph kites; see under that species for distinctions.

(5) **Gray-headed Kite** (plate 4) white-morph juveniles are similar to juvenile male Hook-bills, but they are streaked, not barred, on underparts.

(6) **Ornate Hawk-Eagle** (plate 25) juvenile is similar to juvenile Hook-bills; see under that species for distinctions.

(7) **Collared Forest-Falcon** (plate 28) light adults also show a white hind collar like those of juvenile Hook-bills; see under that species for distinctions.

STATUS AND DISTRIBUTION Uncommon breeding residents on both slopes, from the US border on the Caribbean slope and c Mexico on the Pacific slope to Panama, but absent on the Mexican plateau and high mountains from s Mexico to Nicaragua. They are migratory in n Mexico and are counted in numbers at Veracruz, Mexico, but in much larger numbers in late migration in s Belize.

HABITAT Occurs in a variety of forests, from dry thorn forest to rain forest, that have high densities of snails. Also occurs in forests near water bodies that have snails.

BEHAVIOR They eat land snails, which they locate by hunting from a perch in thick forest. Their favorite places for snail extraction are marked by piles of snail shells on the ground below. They soar for a while on early-morning thermals and have been reported soaring in a flock.

Hook-billed Kite. Adult female Dark. Dark adult males are similar but more glossy black. Texas USA. July 2002.
WSC

Call is a distinctive rattling, uttered when the bird is disturbed near its nest and during courtship.

Active flight is with slow, rather deep, and languid wing beats of bowed wings, with wrists slightly cocked up and wingtips down. Soars and glides with wings slightly bowed, usually not very high, and often just above treetops. Courtship flight is with exaggerated slow, deep wing beats and is performed by both sexes and often a third kite, presumably last year's young.

MOLT Not well known. Apparently annual molt is complete for adults, beginning in spring and continuing through autumn. Second prebasic molt of flight and tail feathers most likely begins in late spring of the second year, although the pre-formative molt of body of some (all?) begins in late autumn with the head and progresses downward.

DESCRIPTION Sexes are similar in size, but have different adult plumages. Juvenile plumage is somewhat similar to that of adult female. Dark-morph kites also have distinct plumages for adult males and females and juveniles. Head shows distinctive greenish lores and bare yellow-orange spots above the eyes. Short legs are dull yellow to yellow. Small feet are rather weak. **Adult male.** Head, back, and upperwing coverts are slate gray. They do not show a pale collar. Underparts are slate gray, with white barring, more noticeable on the belly. Dark gray underwings show bold white barring on primaries. Slate-gray tail usually has one wide band visible, which appears whitish to pale gray above and white below; usually a second white band is visible at tail base. Adult eyes are whitish. **Adult female.** Head has gray face, dark gray crown and nape, rufous cheeks, and creamy throat finely barred rufous. Note distinctive, wide, buffy-rufous collar on hind neck. Back and upperwing coverts are dark brown, sometimes with a grayish cast. Buffy underparts show wide rufous-brown barring. Pale underwings show rufous and buffy barred coverts and heavily banded primaries and outer secondaries; inner primaries have a rufous cast. Dark brown tail has two wide bands, which appear pale gray-brown above and white below. **Juvenile.** Similar to adult female, but crown and nape are blackish, eyes are medium brown, cere is yellowish, collar on hind neck is white, and whitish underparts have a variable amount of narrow dark brown barring, from sparse (males) to heavy (female). Dark brown back and upperwing covert feathers have buffy to rufous edgings. Pale underwings show creamy coverts and heavily barred primaries and outer secondaries, and lack any rufous wash. Tail shows three dark and light bands of equal width. **Juveniles in molt.** They can show adult body but juvenile flight feathers. **Dark-morph adult.** Entire body and coverts are dark slate gray (male) or blackish-gray (females). Flight feathers are black, but some can show a few whitish spots on outer primaries. Black tail has one wide white band. **Dark-morph juvenile.** Entire body and coverts are brownish black. Outer primaries are barred black and white. Dark tail has two wide white bands.

FINE POINTS In parts of its range, this species occurs in two distinct forms that differ only in beak size; one form has a huge beak, thought to be larger because they eat larger snails. Large-beaked kites have been reported from parts of Mexico.

UNUSUAL PLUMAGES No unusual plumages have been reported.

HYBRIDS No hybrids have been reported.

ETYMOLOGY *Chondroheirax* is from the Greek *chondros,* "composed of cartilage," and *hierakos,* "falcon or hawk"; *uncinatus* is Latin for "hooked."

Hook-billed Kite. Juvenile. Juvenile males are sparsely barred on underparts. Juvenile females show heavier dark barring. Panama. November 2010.
JOHN AFDEM

SWALLOW-TAILED KITE PLATE 5

Elanoides forficatus

ELANIO TIJERETA

LENGTH 52–62 cm **WINGSPREAD** 119–136 cm **WEIGHT** 325–500 g

Swallow-tailed Kite.
Adult. Note lack of
a supra-orbital ridge
and blackish mantle.
Florida. April 2012.
KEVAN SUNDERLAND

IDENTIFICATION SUMMARY Graceful flyers and unmistakable with their bold black and white plumage; long, pointed wings; and long, deeply forked tail.

TAXONOMY AND GEOGRAPHIC VARIATION *E. f. yetapa* breeds throughout locally. The North American race, *E. f. forficatus*, occurs as a migrant in spring and autumn. But the subspecies hardly differ.

SIMILAR SPECIES No other North or Central American raptor is similar.

STATUS AND DISTRIBUTION Local and uncommon breeders from se Mexico to Colombia, primarily on the Caribbean slope north of

Nicaragua. All but a few kites migrate into South America during the northern winter. The kites from the United States migrate through Mexico and Central America in autumn and spring, mainly on the Caribbean slope; most pass through the Yucatán Peninsula.

HABITAT Occurs above a variety of humid and pine forests, more commonly in foothills and mountains. Migrants occur in a wider variety of habitats, including less-humid forests.

BEHAVIOR Graceful flyers, they spend much time in flight hunting, riding air currents on steady wings with hardly a wing beat, using the tail as a rudder. Active flight is with slow, flexible wing beats. Soars and usually glides on flat wings, but sometimes glides with wrists below the body and wingtips up. They hunt mainly on the wing for a variety of prey: flying insects, tree frogs, hymenopteran nests, bats, lizards, and snakes, all of which they deftly pluck from the ground, air, or tree branches. They also snatch bird nests and extract the nestlings. They have also been reported to eat fruit.

Swallow-tailed Kite. Adult. All often show rufous staining on belly and undertail coverts. French Guinea. January 2014.
MICHEL GIRAUD-AUDINE

113

Kites often eat in flight, bending the head down to bite off a morsel from prey held in the feet, and drink on the wing in a swallow-like manner.

They build a smallish stick nest high in trees, often communally.

During courtship flights, pairs fly in close formation, calling "klee, klee, klee." Swallow-tailed Kites are especially social; they are often seen in groups of up to 50 individuals and form communal night roosts.

MOLT Not well studied. Many adults begin annual molt in summer before breeding, but apparently complete their molt after breeding. Migrants suspend molt white traveling and finish on the winter grounds in South America. Juveniles in the northern spring have usually undergone no molt; they most likely begin the molt into adult plumage later that summer and complete it on their winter grounds.

DESCRIPTION Sexes are alike in plumage, but females average slightly larger. Plumage of juvenile is similar to that of adult. Cere is dull blue-gray, and short legs are gray. **Adult.** Head, underparts, underwing coverts, undertail coverts, and a small area, usually covered, on lower back above rump are pure white. Undertail coverts and belly sometimes appear soiled. Upper back and lesser and median upperwing coverts are black, and lower back, rump, greater upperwing coverts, flight feathers, uppertail coverts, and tail are blue-black and show a purplish to bronze iridescence in good light. In flight, upperwings and back show a silvery sheen (apparently because of a film of powder down on feathers). Some scapulars are mostly white, visible as white patches on back of perched birds, but are usually not seen on flying kite. Eyes are dark brown. **Juvenile.** Similar to adult, but black areas have more greenish, less purplish, iridescence, and some upperwing coverts and flight and tail feathers have narrow white tips, which usually wear off by winter. Forked tail is shorter than that of adult. Fine black shaft streaks on head, nape, and breast feathers are usually not noticeable in the field. Fledglings have rufous bloom on upper breast that disappears soon after fledging. Eyes are paler brown.

FINE POINTS Swallow-tailed Kites lack bony projections above the eyes (supra-orbital ridge) and so have a pigeon-headed appearance. All have dark brown eyes, except for four in captivity that had orange to red eyes. Red eyes are reported in many references, perhaps erroneously.

UNUSUAL PLUMAGES No unusual plumages have been reported.

HYBRIDS No hybrids have been reported.

ETYMOLOGY *Elanoides* is from the Greek *elanos*, "kite," and *oideos*, "resembling"; *forficatus* in Latin means "deeply forked." The AOU changed the name from American Swallow-tailed Kite to Swallow-tailed Kite in 1996.

Swallow-tailed Kite. Juvenile. Juveniles have shorter forked tails. Mexico. September 2005. ANTONIO HIDALGO

PEARL KITE PLATE 6
Gampsonyx swainsonii
ELANIO CHICO

LENGTH 20–25 cm **WINGSPAN** 45–55 cm **WEIGHT** 74–105 g

IDENTIFICATION SUMMARY Small, pale, falcon-like kites. Patterned head shows white forehead, cheeks, and hind collar and dark rear crown. Note dark semi-collar on upper breast. Body, wings, and tail are dark above and whitish below. Wingtips appear pointed on flying kites and fall short of tail tip on perched kites.

TAXONOMY AND GEOGRAPHIC VARIATION *G. s. leonae* occurs throughout Central America, with no geographic variation.

SIMILAR SPECIES (1) **American Kestrel** (plate 29) has a similar silhouette in flight and also perches on wires beside roads, but has much different plumage; especially, the head has two black mustache marks (vertical bars), and the underparts are marked with streaking or spotting.

Pearl Kite. Adult. Adults have rufous flanks. Ecuador. March 2013.
NICK ATHANAS

STATUS AND DISTRIBUTION Uncommon to fairly common, but local, from Panama to El Salvador on the Pacific slope. They are slowly expanding in range northward.

HABITAT Occurs in a variety of open habitats, including in savannahs and parklands, along roads through forest, and along rivers in arid habitats.

BEHAVIOR They are most often seen perched on conspicuous perches, kestrel-like, searching for prey of insects and small lizards.

Powered flight is quick and direct. They soar and glide with wings held level. They may hover momentarily before pouncing on prey.

MOLT Not studied. Presumably juveniles molt into adult plumage within a year of fledging.

DESCRIPTION A small, pale, falcon-like kite. The patterned head shows a white forehead, cheeks, and hind collar and a dark rear crown. Wingtips appear pointed on flying kites and fall short of the tail tip on perched kites. Cere is blue-gray.

Adult. Crown and upperparts are sooty-gray. Forehead and cheeks are pale

cinnamon. Creamy underparts show rufous flanks, and leg feathers are pale cinnamon. Note narrow, dark semi-collar on sides of upper breast and pale hind collar with narrow rufous band at bottom. Remiges and rectrices are blackish above and unmarked silvery below. Underwing coverts are pale cinnamon. Eyes are red, and legs are orange-yellow. **Juvenile.** Almost like adult, but with darker gray on crown and upperparts, with faint, narrow, pale feather edging on upperparts and a cinnamon wash on underparts in fresh plumage. Hind collar is wider than adult's and lacks rufous at bottom. They have white tips on primaries, brown eyes, and yellow legs, and they lack rufous on flanks.

FINE POINTS They were originally thought to be a falcon, but their molt is accipitrid and not falconid.

UNUSUAL PLUMAGES No unusual plumages have been described.

HYBRIDS No hybrids have been described.

ETYMOLOGY *Gampsonyx* comes from the Greek *gampos* and *onux*, "curved" and "claw." The species name is after William Swainson, an English artist, collector, and author.

Pearl Kite. Juvenile. Juveniles have darker upperparts and lack rufous flanks. Brazil, June 2006. LOU HEGEDUS

WHITE-TAILED KITE PLATE 5
Elanus leucurus

ELANIO MAROMERO

LENGTH 36–41 cm **WINGSPREAD** 99–102 cm **WEIGHT** 305–361 g

IDENTIFICATION SUMMARY Falcon-like in shape and gull-like in color and flight. Charcoal-black upperwing coverts form the black shoulders on perched birds and black carpal patches on underwings of flying birds. Their stoop for prey is distinctive: lowering slowly (parachuting) with wings held in a deep V high above the back and legs dangling, and finishing with a rush to the ground. Tail-cocking behavior—the tail raised repeatedly over the back—is also distinctive; no other American raptor does this. Wingtips almost reach tail tip on perched kites.

TAXONOMY AND GEOGRAPHIC VARIATION The White-tailed Kite was regarded for a few years in the 1980s as a race of the Black-shouldered Kite (*Elanus caeruleus*) based on inadequate information, but it is now again classified by the AOU as a separate species (see Clark and Banks 1992). The nominate subspecies of South America occurs in Panama and *E. c. majusculus* occurs north of there, but they have the same plumage, and the northern kites are slightly larger.

SIMILAR SPECIES (1) **Mississippi Kite** (plate 8) has dark body and tail in all plumages.

(2) **Northern Harrier** (plate 9) adult male appears somewhat similar; see under that species for distinctions.

(3) **Falcons** lack white unbarred tail and black carpal patches.

(4) **Adult gulls** have longer necks and a shorter tail and lack black carpal patches on underwings.

STATUS AND DISTRIBUTION Fairly common to common, but local, and widespread throughout in open areas. In Panama they are more common on the Pacific slope. Their numbers throughout have increased greatly since the 1950s.

HABITAT Occurs in a variety of open habitats, especially grasslands, marshes, open savannahs, rice fields, and cleared areas.

BEHAVIOR They behave somewhat like large American Kestrels, hovering over open areas looking for prey. Prey is almost exclusively rodents, which they always hunt by hovering. As they plunge after prey, they hold their wings fully stretched upward. At times they are more active just after sunrise and before sundown.

Active flight is with light, steady wing beats of cupped wings. Soars with wings in a medium dihedral with tips somewhat rounded. Glides with wings in a modified dihedral. Hovers regularly, sometimes with legs dangling. Display flight of adult male is with fluttering wings held in a deep V above the body and is accompanied by raspy,

creaking calls. Adults have been observed cartwheeling. A few to several hundred kites form communal night roosts in fall and winter.

Two usual vocalizations are a soft, sweet whistle, used only when breeding, and a harsh, raspy alarm call.

MOLT Annual molt of adults is complete, beginning in spring and completed by autumn, but may be suspended during parts of the breeding cycle, especially by the male. Preformative molt begins approximately two to three months after fledging, continues for another three months or more, and is a more or less complete molt of body feathers; it sometimes includes a few tail feathers. Flight and the rest of the tail feathers are not replaced until the next spring and summer. Formative plumage kites have adult body and tail coverts, but retain juvenile flight feathers and wing coverts, with white tips of the latter usually visible as narrow pale lines on both flying and perched kites. Tail molt precedes wing molt. Molt varies with fledging date; kites from early broods show more molt.

DESCRIPTION Sexes are almost alike in plumage and size. Immature plumages are similar to that of adults. Cere and legs are yellow. **Adult.** White head has narrow black circles around and small black areas in front of eyes. Eyes are orange-red to scarlet. Crown, nape, back, primary and greater secondary upperwing coverts, and uppersides of secondaries are medium gray (males average paler), and uppersides of primaries are a bit darker gray. Lesser and median upperwing coverts are charcoal black. Underparts are white. Underwing shows blackish primaries, off-white to whitish secondaries, and white coverts, with a small black carpal patch. Tail is white, except for gray central feathers. **Juvenile.** Somewhat similar to adult, but eyes are brown, lightening with age. Gray-brown rear crown, nape, and back feathers have whitish edgings. Black upperwing coverts show

OPPOSITE PAGE: White-tailed Kite. Adult pair. Adult males have white fore-crowns and females have gray fore-crowns. California USA. December 2010.
RICHARD PAVEK

ABOVE: White-tailed Kite. Adult pair. Adult males have white fore-crowns and females have gray fore-crowns. California USA. May 2010.
RICHARD PAVEK

119

White-tailed Kite.
Juvenile. Juveniles just
fledged show brownish
gray crown and
upperparts and rufous
on breast. Note faint
gray subterminal band.
California USA.
February 2011.
RICHARD PAVEK

buffy edgings. Underparts are white with some narrow dark streaking or spotting and a rufous wash across breast that fades or is molted soon after fledging. Gray flight feathers and greater upperwing coverts have white tips. Undersides of secondaries are pale gray and contrast somewhat with white wing linings. White tail feathers have a narrow gray subterminal band. Juveniles undergo a more or less complete post-juvenile molt of body feathers (sometimes a few tail feathers) within months after fledging. **Formative plumage**. Older immatures in their first winter after undergoing preformative molt appear much more adult-like, but are distinguished by white tips to greater upperwing coverts and flight feathers and narrow dusky subterminal tail band. Eyes are pale orangish to orange.

FINE POINTS A few kites lack black carpal patches on underwings. Adults and juveniles have pale gray central tail feathers, and folded uppertails appear this color on perched birds; from below the tail appears white. Central and outer tail feathers are noticeably shorter than the others.

UNUSUAL PLUMAGES No unusual plumages have been described.

HYBRIDS No hybrids have been reported.

ETYMOLOGY *Elanus* is Latin for "kite"; *leucurus* is Greek for "white-tailed."

REFERENCES Clark and Banks 1992.

SNAIL KITE PLATE 7

Rostrhamus sociabilis

CARACOLERO COMÚN

LENGTH 41–47 cm   **WINGSPAN** 104–112 cm   **WEIGHT** 340–520 g

IDENTIFICATION SUMMARY Gregarious kites, typically found near water. They have a thin, deeply hooked beak, white tail coverts and tail base, and paddle-shaped wings shown when flying. Sexes are similar in size, but have different adult plumages. Juvenile and Basic

Snail Kite. Adult male. Adult males are gray overall. Note long curved beak. Belize. December 2004. wsc

121

II (second) plumages are similar to that of adult female. Dark tail has square tip and white base; tail coverts are white. On perched kites, wingtips extend beyond tail tip.

TAXONOMY AND GEOGRAPHIC VARIATION Two subspecies have disjunct ranges, but differ only in size, with northern *R. s. major* being larger than southern nominate kites.

SIMILAR SPECIES: (1) **Northern Harrier** (plate 9) also has white uppertail coverts, but has longer, narrower wings, flies in a more direct manner with wings in a dihedral, and has white uppertail coverts, but no white on tail.

(2) **Slender-billed Kite** (plate 7) is similar to adult males; see under that species for distinctions.

(3) **Hook-billed Kite** (plate 4) is somewhat similar in having paddle-shaped wings and similar plumages, but never shows orange facial area and legs.

(4) **Osprey** (plate 3), when perched, shows a similar head pattern to faded immature Snail Kite; see under that species for distinctions.

STATUS AND DISTRIBUTION Fairly common but local, restricted to freshwater marshes and other wet areas where their favored prey of apple snails occur. Nominate kites occur from Panama north, and *R. s. major* from s Mexico to n Honduras, but only on the Caribbean slope.

Snail Kite. Juvenile. Juveniles are similar to adult females but with buffy markings on face and buffy underparts. Belize. December 2012. WSC

HABITAT Occurs in a variety of freshwater habitats, including marshes, rice fields, and ponds.

BEHAVIOR These highly social kites are often seen in groups. They feed almost exclusively on snails of the genus *Pomacea*, which they deftly pluck from the water surface. They forage by flying slowly above the water with more or less continuous slow flapping at a height of two to ten meters. When a snail is sighted, they brake, swoop down, and pluck the snail from the water with a foot, then transfer the snail from foot to beak in the air. Kites also hunt from perches, and they perch to extract snails from shells using their specialized beak. Hunting can take place at any time of day. Some prey items other than snails have been reported. Nesting and night roosting are usually colonial. During the middle of the day, Snail Kites often soar, sometimes high enough to be out of sight from the ground.

Active flight is buoyant, with slow, floppy wing beats of cupped wings. Glides on cupped wings, and soars with wings somewhat less cupped.

They build a bulky nest in a bush or small tree near water.

Vocalizations are a series of guttural calls.

MOLT Annual molt is complete.

DESCRIPTION They have a thin, deeply hooked beak, white base of tail, and white uppertail coverts, and show paddle-shaped wings when flying. Sexes are similar in size, but have different adult plumages. Juvenile and Basic II (second) plumages are similar to that of adult female. Dark tail has square tip and white base; tail coverts are white. On perched kites, wingtips extend beyond tail tip. **Adult male.** Entire body and wing coverts are slate gray, darker on back, head, and upperwing. Iris color is carmine, and cere and face skin are bright red-orange. Flight feathers are slate-black. Legs are bright orange. **Adult female.** Head has dark brown crown and nape, and a distinctive face pattern of buffy superciliary line, dark eye-line, and buffy cheeks and throat. Cere and face skin are yellow to orange. Eyes are dark carmine. Back and upperwing coverts are dark brown, with

Snail Kite. Juvenile. All glide with paddle-shaped wings bowed—wrists up and tips down. Venezuela. December 2005. WSC

rufous feather edges. Creamy underparts have heavy, irregular dark brown streaking. Underside of wing shows dark brown coverts and barred flight feathers, usually with a light patch at base of outer primaries. Legs are yellow-orange. Older females look somewhat like adult males, with a blackish head that lacks the buffy superciliary line, blackish nape and upper back, and mostly dark underparts with some whitish mottling, but they retain an overall brown cast and always show remiges that are narrowly banded and paler than an adult male's. **Juvenile.** Similar to adult female, but eye color is dark brown, cere is whitish to pale yellow, top of buffy head is streaked, buffy underparts are streaked, tail is browner, and legs are yellow. **Basic II (second plumage) kites.** Similar to juveniles, but have whiter and more heavily marked heads, whiter underparts, and yellow-orange legs and face skin. **Basic III male.** Similar to adult males, but with heavily spotted undersides of dark remiges.

FINE POINTS Basic III males (two to three years old) are in transition to adult plumage and are grayer and uniformly darker than adult females. In times of plenty, kites may breed in juvenile plumage.

UNUSUAL PLUMAGES No unusual plumages have been reported.

HYBRIDS No hybrids have been reported.

ETYMOLOGY *Rostrhamus* is from the Latin *rostrum*, "beak," and the Greek *hamus*, "hook;" *sociabilis* is Latin for "gregarious." The Florida race was formerly called the Everglade Kite. The common name is from its preferred prey of snails.

SLENDER-BILLED KITE PLATE 7

Helicolestes hamatus

CARACOLERO CHICO

LENGTH 35–41 cm **WINGSPAN** 80–90 cm **WEIGHT** 375–485 g

IDENTIFICATION SUMMARY Darkish gray kites usually found near water. Vagrant to eastern Panama. Plumage of adults is alike by sex; juvenile plumage is similar. Eyes are whitish. Legs, cere, gape, and face skin are bright orange-red. Wings are pushed forward when soaring. All glide with wings heavily bowed. Short tail is square-tipped. Wingtips reach tip of short tail on perched kites.

TAXONOMY AND GEOGRAPHIC VARIATION Monotypic, with no geographic variation. Formerly included in the genus *Rostrhamus*, but DNA has shown that they are not close to Snail Kites.

SIMILAR SPECIES (1) **Snail Kite** (plate 7) adult male is also overall dark gray, but has scarlet eyes and longer wings and tail, shows white on tail coverts and base of tail, and does not push wings forward when soaring. Wingtips of perched Snail Kites extend beyond tail tip. (2) **Crane Hawk** (plate 9) in Panama is also overall gray like adults, but they are much slimmer, have dark eyes and a much longer tail and legs, and lack the long narrow curved beak.

Slender-billed Kite. Adult. Note lack of white in plumage and wingtips reach tip of short tail. Colombia. March 2008.

NICK ATHANAS

STATUS AND DISTRIBUTION Vagrants or rare local residents in e Panama (Darien Province).

HABITAT Occurs in lowland wet forests, especially at edges of rivers and lagoons.

BEHAVIOR Unobtrusive still hunters that perch in waterside trees and shrubs, from which they launch short flights to capture apple snails. They are usually found singly or in pairs, never in groups.

They soar occasionally, but usually not high, and glide with wings cupped: wrists up and tips down. Active flight is with slow, floppy wing beats.

Vocalizations are soft mewing calls.

MOLT Not studied. Apparently, annual molts are complete.

DESCRIPTION Eyes are whitish to dull yellow. Wingtips reach tip of short tail. **Adult.** Sexes are alike. They are overall unmarked darkish gray, except for black tips of outer primaries. Legs, cere, gape, and face skin are bright orange-red. Short blackish tail is square-tipped. **Juvenile.** Similar to adult in being overall darkish gray, but undersides of remiges are heavily banded white, tail shows white bands, and upperparts and leg feathers show brownish feather edges. Cere, gape, and face skin are bright yellow-orange to orange.

UNUSUAL PLUMAGES No unusual plumages have been described.

HYBRIDS No hybrids have been described.

ETYMOLOGY *Helicolestes* comes from the Greek *helix*, "twisted," as in a snail shell, and *lestes*, "robber." The species name, *hamatus*, is from the Latin *hamus* for "hooked" or "curved," in reference to its beak.

Slender-billed Kite. Juvenile. Juveniles are similar to adults but have white banding in remiges and white tail bands. Suriname. May 2014.
JEAN-LOUIS ROUSSELLE

127

DOUBLE-TOOTHED KITE PLATE 6
Harpagus bidentatus

MILANO BIDENTADO

LENGTH 32–37 cm **WINGSPAN** 60–72 cm **WEIGHT** 159–238 g

IDENTIFICATION SUMMARY Small, accipiter-like kites. They soar and glide with wings below the horizontal and tail folded, often with undertail coverts widely spread. Tail above appears dark with three narrow whitish bands, and below shows wide, even-width black and dusky bands, with narrow white bands in the latter. Adult male, adult female, and juvenile have different plumages. All show dark throat stripe. Ceres are dull green or yellow, and legs are yellow. Wingtips fall somewhat short of tail tip on perched kites.

TAXONOMY AND GEOGRAPHIC VARIATION *H. b. fasciatus* occurs from Mexico to Panama and into w South America, with no geographic variation in our area.

SIMILAR SPECIES (1) **Sharp-shinned Hawk** (plate 10) in flight is similar in size and shape, but usually spreads tail, has rufous or brown underwing coverts and narrower pale bands on tail, and does not flare the undertail coverts. Perched Sharpies lack dark mesial stripes.

(2) **Bicolored Hawk** (plate 11) is somewhat similar in also being an accipiter, but is larger and always shows unmarked underparts and narrower pale tail bands.

Double-toothed Kite. Adult male. Adult males have gray barring on the breast. All show dark throat stripe. Mexico. October 2015.

GREG LASLEY

(3) **Barred Forest-Falcon** (plate 28) adults are similar; see under that species for distinctions.

STATUS AND DISTRIBUTION Fairly common in a variety of wetter forests from s Mexico to Colombia, more common on the Caribbean coast.

HABITAT Occurs in humid and wet forests and well-watered, treed savannahs, second growth, and other open woodlands.

BEHAVIOR These still hunters perch on forest edge or canopy. They prey on insects and lizards that they capture in trees or on the ground after a short stoop. They also capture insects in the air after a short flight. They regularly follow monkey groups.

They soar with wings held level or slightly depressed, often with their

undertail coverts flared, and they glide with their wings depressed, both usually with the tail folded. Powered flight is accipiter-like, with three to five deep wing beats followed by a short glide. Display flights described include mutual soaring on depressed wings and undulating flight, with three to five rapid wing beats on the upsweep, both with undertail coverts flared.

They build small compact stick nests in a tall tree, and their calls are high and thin and not very raptor-like; to some, the calls are more like those of tyrant flycatchers.

MOLT Not well studied. Annual molts are complete. Juveniles most likely molt into adult plumage by the end of their first year.

DESCRIPTION A small, accipiter-like kite. The plumages of adult male, adult female, and juvenile differ. All show dark throat stripe. Tail above appears dark with three narrow whitish bands, and below shows wide even-width black and dusky bands, with narrow white bands in the latter. Ceres are dull yellow to green, and short legs are yellow to yellow-orange. Adults' eyes are orange, and juveniles' eyes are yellow. **Adult male.** Blue-gray head shows white throat with noticeable dark central stripe. Upperparts are blue-gray. Whitish underparts are boldly barred gray, with a rufous wash on sides of upper breast. **Adult female.** Similar to adult male, but somewhat larger; breast is uniformly rufous, and belly is barred rufous. **Juvenile.** Brown head shows white throat with noticeable dark central stripe. Upperparts are dark brown. Whitish underparts lightly streaked dark brown (males) or more heavily streaked (females).

FINE POINTS Sharp-shinned Hawks appear similar in soaring flight, but they usually soar with tail flared, whereas Double-tooths keep their tail closed while soaring.

UNUSUAL PLUMAGES No unusual plumages have been described.

HYBRIDS No hybrids have been described.

ETYMOLOGY *Harpagus* comes from the Greek *harpazo*, "to seize or plunder." The species name, *bidentatus*, is Latin for "having two teeth"; this and the common name refer to undulations in the edges of the mandibles.

ABOVE LEFT: Double-toothed Kite. Adult. Note fluffed white undertail coverts and tail held folded. Mexico. April 2014. GREG LASLEY

ABOVE RIGHT: Double-toothed Kite. Juvenile. Juveniles have marked whitish underparts, heavier on females. Belize. August 2015. RYAN PHILLIPS—BRRI

MISSISSIPPI KITE PLATE 8

Ictinia mississippiensis

MILANO MISSISSIPPI

LENGTH 31–37 cm **WINGSPREAD** 75–83 cm **WEIGHT** 240–372 g

IDENTIFICATION SUMMARY Dark, falcon-shaped aerial hunters seen in our area only on migration in spring and fall. The sexes are nearly alike in adult plumages that differ mainly in the undertail coverts and the undertail, but the juvenile plumage is quite different. Juveniles returning in the spring (Basic II, second plumage) have adult-like gray bodies, but have retained juvenile tail and flight feathers. Outermost primary is noticeably shorter than others. Wingtips of perched birds reach or extend barely beyond tail tip.

TAXONOMY AND GEOGRAPHIC VARIATION Monotypic, with no geographic variation.

SIMILAR SPECIES (1) **Plumbeous Kite** (plate 8) is a similar overall gray, pointed-wing kite; see under that species for distinctions.

(2) **Peregrine Falcon** (plate 31) is similarly shaped, but has dark head and a malar stripe, and lacks flared tail. Peregrine's flight is more powerful and purposeful, not leisurely and buoyant.

(3) **White-tailed Kite** (plate 5) always has white body and tail and black carpal mark on underwings.

STATUS AND DISTRIBUTION The entire population passes through e and s Mexico and Central America on migration in both spring and fall. Migration path in autumn is along the Caribbean coast of Mexico until well south, then along both slopes down to South America,

BELOW LEFT: Mississippi Kite. Adult male. Adult males have gray undertail coverts and all black undersides of tail feathers. Kansas USA. August 1994. WSC

BELOW RIGHT: Mississippi Kite. Adult female. Adult females have whitish undertail coverts and pale underside of outer tail feathers with a dusky tail tip. Texas USA. April 2015. WSC

although more common on the Caribbean slope. Path is reversed in spring. They bypass the Yucatán Peninsula, but do occur occasionally in Belize along the coast.

HABITAT They fly over a variety of habitats as they migrate through our area.

BEHAVIOR They are aerial hunters, spending much time on the wing. Their main prey, insects, are caught and often eaten on the wing.

Their active flight is leisurely, on flexible wings. Flight is light, buoyant, and effortless. They soar and glide on flat wings, rarely flapping. Wingtips often curl up during soaring.

They are gregarious and migrate in flocks of from several to several hundred individuals and roost together at night. They are usually silent away from the breeding areas in the US.

MOLT Annual molt of adults is apparently complete; most of it done in winter in South America. Juveniles initiate body and covert molt during their first winter and return in spring appearing somewhat adult-like in Basic II (second) plumage. Molt of the remiges and rectrices begins after their arrival on the breeding grounds, is suspended during the autumn migration, and is completed on the winter grounds.

DESCRIPTION Sexes are similar in size and differ slightly in plumage. Juvenile plumage is different. Juveniles returning in spring are intermediate in plumage. **Adult.** Head is pale gray, darker on females, with a small area of black in front of the eye and a narrow ring

131

around the eye. Back and upperwing coverts are slate gray. Pale gray uppersides of secondaries form a pale bar on the sides of perched birds and a wide whitish band on the inner trailing edge of the upperwing on flying birds. Underparts are medium gray. Underwings are slate gray with a short, narrow whitish band on the inner trailing edge. Males have some rufous on inner primaries and outer secondaries, sometimes noticeable on flying kites; rufous is more restricted on female. Undertail coverts of females are gray with some whitish barring. Flared tail is all black on males, but undersides of outer tail feathers of females are pale with a darker terminal band. Short legs are orange-yellow with a dusky cast on the fronts and the upper toes. Cere is dark gray. Eyes are scarlet. **Juveniles.** Head is dark gray-brown, with fine whitish streaks and a short buffy superciliary line. Creamy throat is unstreaked. Back and upperwing coverts are dark brown with rufous feather edging. Underparts are creamy, with a variable amount of dark rufous-brown streaking that ranges from sparse to heavy. Underwing has mottled rufous-brown coverts and somewhat darker flight feathers that show a variable amount of whitish areas. Flight feathers have white tips. Short legs are yellow with a dusky cast on the fronts and the upper toes. Dark brown tail has three incomplete whitish bands that are sometimes absent. Cere is greenish-yellow to yellow. Eyes are dark brown. **Basic II (second plumage).** Juveniles returning after their first spring have molted into an adult-like gray head and body that often shows small, oval white blotches both above and below, the result of retained juvenile feathers and whitish bases to first adult feathers. Flight and tail feathers are retained from juvenile plumage. They initiate molt of remiges and rectrices soon after arrival and, by autumn migration, have replaced inner six or seven primaries and outer and central tail feathers. Cere and eyes as in adults.

FINE POINTS The relatively short outermost primary extends only four-fifths of the way to the wingtip; this can be seen on flying kites.

UNUSUAL PLUMAGES No unusual plumages have been described.

HYBRIDS No hybrids have been reported.

ETYMOLOGY *Ictinia* is from the Greek *iktinos*, "kite." "Mississippi" and *mississippiensis* refer to the state in which the type specimen was collected by Wilson. Species name was originally spelled *misisippiensis*, a printer's error. This form was used until the AOU corrected it in 1976.

PLUMBEOUS KITE PLATE 8

Ictinia plumbea

MILANO PLOMIZO

LENGTH 30–38 cm   **WINGSPREAD** 80–97 cm   **WEIGHT** 220–374 g

IDENTIFICATION SUMMARY Dark, falcon-shaped, aerial hunters. Sexes are alike in adult plumage, but juvenile plumage differs. Juveniles returning in the spring (Basic II, second plumage) have adult-like gray bodies, but have retained juvenile tail and flight feathers. Outer primary is noticeably shorter than others. Wingtips of perched birds extend far beyond tail tip.

TAXONOMY AND GEOGRAPHIC VARIATION Monotypic, with no geographic variation.

SIMILAR SPECIES (1) **Mississippi Kite** (plate 8) appears similar in shape, flight, and plumages, but they have proportionally shorter wings and, on perched kites, wingtips reach or barely exceed tail tip. Adults are overall a paler gray, especially their heads, and show much less rufous in the primaries from below and no rufous from above. Juveniles are streaked rufous-brown on underparts and underwing coverts, have dark brown eyes, and seldom show rufous in the primaries.

(2) **White-tailed Kite** (plate 5) always has white body and tail and black carpal mark on underwings.

(3) **Peregrine Falcon** (plate 31) is similarly shaped, but has dark head, a malar stripe, and lacks flared tail. Peregrine's flight is more powerful and purposeful, not leisurely and buoyant.

STATUS AND DISTRIBUTION Uncommon to fairly common, but local in some areas, as breeders from s Mexico to Panama on both slopes. Almost all migrate into South America for the northern winter.

FAR LEFT: Plumbeous Kite. Adult. All show wingtips extending way beyond tail tip. Adults are overall gray with white tail bands. Belize. March 2010.
RYAN PHILLIPS—BRRI

LEFT: Plumbeous Kite. Adult. Adults in flight show rufous remiges and two white tail bands. Belize. March 2007.
RYAN PHILLIPS—BRRI

133

Plumbeous Kite. Juvenile. Juveniles have heavily streaked underparts. Costa Rica. June 2013. NICK ATHANAS

HABITAT Occurs in a variety of forests, usually more open ones, and more commonly in hilly country. They also forage in adjacent open areas.

BEHAVIOR They are aerial hunters, spending much time on the wing, usually flying above forests. Their main prey, insects, are caught and often eaten on the wing. They also capture other prey, such as lizards, snakes, and frogs from trees, when breeding. They often perch on high snags.

Active flight is leisurely, on flexible wings, and is light, buoyant, and effortless. They soar and glide on flat wings, rarely flapping. Wingtips often curl up during soaring.

They build a small stick nest in tall trees, and they are usually silent, except when breeding, when they utter whistled calls.

MOLT Annual molt of adults is apparently complete; most of it done in winter in South America. Juveniles initiate body and covert molt during their first winter and return in spring, appearing somewhat adult-like in their second (Basic II) plumage. Molt of adults' remiges and rectrices begins after their arrival on the breeding grounds, but this is suspended while breeding and during the autumn migration and completed on the winter grounds.

DESCRIPTION Sexes are similar in size and plumage. **Adult.** Head shows a small area of black in front of the eye and a narrow ring around the eye. Cere is dark gray. Eyes are scarlet. Back and upperwing coverts are dark gray; head and underparts are somewhat paler, but still dark. Primaries are rufous, with dark gray areas on uppersides. Secondaries are dark gray. Short legs are orange. Dark tail is black and shows two white bands. **Basic II (second plumage).** Juveniles returning their first spring have molted into adult-like gray head and body that often show small, oval white blotches both above and below, the result of retained juvenile feathers and whitish bases to the first adult feathers. Eyes are orangish. Cere is dark gray. Flight and tail feathers are retained from juvenile plumage. They initiate molt of remiges and rectrices soon after arrival in spring and, by autumn migration, are in adult plumage. Legs are orangish. **Juveniles.** Head is dark gray-brown, with fine whitish streaks and a short buffy superciliary line. Cere is yellow. Creamy throat is unstreaked. Eyes are pale brown. Back and upperwing coverts are blackish-brown with whitish feather edging. Underparts are creamy, with a variable amount of slate-gray streaking. Underwings show grayish coverts, mostly rufous or mostly whitish primaries, and dark gray secondaries. Remiges have white tips. Short legs are yellow. Dark tail has three whitish bands.

FINE POINTS The relatively short outer primary extends only four-fifths of the way to the wingtip; this can be seen on flying kites.

UNUSUAL PLUMAGES No unusual plumages have been described.

HYBRIDS No hybrids have been reported.

ETYMOLOGY *Ictinia* is from Greek *iktinos*, "kite." "Plumbeous" and *plumbea* are from the Latin *plumbeus*, "lead colored."

BALD EAGLE PLATE 3
Haliaeetus leucocephalus

ÁGUILA CALVA, ÁGUILA CABECIBLANC

LENGTH 70–90 cm **WINGSPREAD** 180–225 cm **WEIGHT** 2.0–6.2 kg

IDENTIFICATION SUMMARY Very large, long-winged raptors, usually found near water. Adults' white head and tail and dark brown body and wings are distinctive. Immatures appear more like Golden Eagles, but always show white "wing pits." In flight, head and neck protrude more than half the tail length. Trailing edge of wing is nearly parallel to leading edge, less so on wings of juveniles.

TAXONOMY AND GEOGRAPHIC VARIATION *H. l. leucocephalus* occurs in n Mexico.

SIMILAR SPECIES (1) **Golden Eagle** (plate 23) in flight has head that protrudes less than half the tail length: head of Bald Eagle protrudes more than half the tail length. Trailing edge of wing is straighter on Balds. Immature Goldens have white on underwing restricted to base of flight feathers; white on Balds is on underwing coverts and axillaries. White axillaries of immature Balds can be seen from a great distance. White in Golden's tail extends to the edges, while immature Balds always have dark edges. Tawny greater upperwing coverts of adult and older immature Goldens form tawny bars on upperwings, visible on flying and perched birds and lacking on all Balds. Perched Goldens show the golden nape and yellow cere and bicolored beak; Balds usually have cere and beak uniformly colored. Bald's tarsi are bare, while Golden's are completely covered with buffy feathers.

(2) **Osprey** (plate 3) in flight shows all-white underparts and black carpal patch on underwings. Their wings are almost always gull-like; Balds never appear so. On perched birds, the small white head with dark eye-line is distinctive.

STATUS AND DISTRIBUTION Rare winter visitor to n Mexico south to Nyarit and n Veracruz and a rare and local resident in Baja California.

HABITAT They are usually found near water where they can find their favorite food, fish. But they can occur in a variety of habitats in the non-breeding season wherever an adequate prey base occurs.

BEHAVIOR This superb fisherman and agile raptor prefers to find food in the easiest possible way. In winter, when fish may be scarce, they also eat carrion and waterfowl. They regularly pirate food from other raptors, especially other Balds and occasionally Ospreys.

Active flight is with slow wing beats, similar to that of Great Blue Heron, but more powerful. They soar usually with wings level or held in a small dihedral. They glide on level wings with wrists cocked forward and wingtips sometimes depressed. They soar often, many times with other raptors.

135

They form communal night roosts, most commonly in winter, of several birds or more and, rarely, several hundred. Individuals often spend considerable time perching. Some birds remain perched in night roosts for a day or two. They are vocal, particularly around other eagles.

MOLT Annual molts of body and flight feathers aren't complete. Second prebasic molt of the body and tail is complete, but not that of the remiges, with only the inner five or six primaries and half of the secondaries replaced; new secondaries are noticeably shorter. Usually S1–S2, S5–S6, and the inner three are new. In the next molt (third prebasic), primary molt begins anew at P1 and continues where it left off after the last molt, with molt proceeding in two places. Usually juvenile P10 is not replaced. All of the juvenile secondaries are usually, but not always, replaced. In the next and subsequent molts, primary molt proceeds in three places or waves. All rectrices are usually replaced annually. Adult plumage is attained after four or five molts, with the first three age classes distinctive, but the fourth and fifth similar.

DESCRIPTION Sexes are alike in all plumages. Legs are orange-yellow. **Adult.** Head, neck, tail coverts, and tail are white, sometimes with a few brown or black spots, even on older birds. Body and wing coverts are dark brown with paler feather fringing. Flight feathers are dark brown. Beak and cere are bright orange-yellow, and eyes are pale lemon-yellow. **Juvenile.** Head is dark brown, sometimes a bit paler on the crown and superciliaries. Beak and

Bald Eagle. Adult. Adults have white heads and tails and orange beaks, ceres, and gapes. British Columbia, Canada. February 2000.
WSC

cere are dark gray, and eyes are dark brown. Back and upperwing coverts are tawny brown and contrast with dark brown flight feathers. Back never shows white triangle of older immatures. Breast is dark brown and usually contrasts with tawny belly, sometimes with short white streaking on underparts. Tail is noticeably longer than in subsequent plumages and can be uniformly dark above, a character not found in later plumages. Undertail is usually dirty whitish with dark edges and dark tip. All immatures in the first three plumages show white axillaries. **Basic II (second plumage).** Brown head shows wide buffy superciliary lines, which contrast with dark cheeks. Eyes vary from light brown to amber, and beak and cere fade to a slate-gray, with cere a bit paler. Dark brown upperwing coverts show white feathers, and dark brown back has a white triangle. Dark brown breast appears as a dark bib that contrasts with variable streaked white belly. Wings now appear ragged on the trailing edge due to a mix of longer pointed juvenile and shorter replacement secondaries, which have dark tips. Tail is dirty whitish with dark edges and dark tip. **Basic III.** Similar to Basic II, but wide superciliary lines and cheeks are now whitish and frame a narrow dark line behind each eye. Eyes are pale whitish-yellow, beak has lightened to horn-colored with some dirty yellow spots, and cere is yellowish. White triangle on back is usually retained, but some eagles this age have mostly dark bellies. Wings are narrower and smooth on the trailing edges, sometimes with one or more longer secondaries protruding. **Basic IV.** Fourth-plumage eagles appear more adult-like. White head and neck usually show Osprey-like dark eye-lines, and white on neck doesn't extend to body as it does on adult. Fore neck and throat are darkish. Eyes are pale yellow, and cere and beak are orange-yellow, with dark smudges on beak. Dark brown body and wing coverts show whitish spotting.

Bald Eagle. Juvenile. Juveniles have tawny bellies and show serrated trailing edge of wings. Virginia USA. January 2001. wsc

Tail is variable, from similar to that of younger immatures to all white with dark tips and streaks. **Basic V.** Some eagles can be fully adult after four molts, but others still show immature characters, such as dark streaks in the white head and dark tips on many tail feathers.

FINE POINTS Bald's wrists during a glide are more forward and primaries are more folded than are those of Golden Eagles in a similar glide. Balds after a series of wing flaps begin gliding with the wings ending on a down stroke, whereas Goldens begin glides ending with an up-stroke.

UNUSUAL PLUMAGES Several amelanistic juveniles and adults have cream-colored body and covert feathers. Partial albinos of all ages have reported.

HYBRIDS Hybrids with Steller's Sea Eagle have occurred, presumably in Alaska, where a mixed pair was observed for many years (Clark 2008).

ETYMOLOGY *Haliaeetus* is from the Greek *halos*, "sea," and *aetos*, "eagle." The species name, *leucocephalus*, is from the Greek *leucos*, "white," and *kephalus*, "head." In Old English, *balde* means "white"; the early Bald-headed Eagle (that is, "White-headed Eagle") later became just Bald Eagle.

REFERENCES Clark 2008.

BLACK-COLLARED HAWK PLATE 14
Busarellus nigricollis

AGUILUCHO ACOLORADO

LENGTH 50–55 cm   **WINGSPAN** 115–135 cm   **WEIGHT** 675–945 g

IDENTIFICATION SUMMARY Distinctive, compact large buteonines, usually found in lowlands near water. White head and wide black necklace are diagnostic. Adults are mostly rufous, juveniles less so. Wingtips reach tail tip on perched birds. Eyes and cere are dark, and legs are gray.

TAXONOMY AND GEOGRAPHIC VARIATION Monotypic, with no geographic variation. Recent DNA studies have shown that this species is more closely related to kites than to buteonines.

SIMILAR SPECIES (1) **Savannah Hawk** (plate 14) adults and second-plumage hawks are similarly overall rufous; see under that species for distinctions.

STATUS AND DISTRIBUTION Rare to fairly common but local on both slopes of Mexico from s Veracruz and s Jalisco to Nicaragua on the Pacific coast and ne Costa Rica on the Caribbean coast, including the Yucatán Peninsula and Belize, and rare and local in ec Panama.

HABITAT Occurs in a variety of lowland wet areas, including marshes, lakes, slow-moving rivers, and coastal mangroves.

BEHAVIOR These still hunters, usually seen perched low on a snag, small bush, or pole overlooking water, prey on fish, but also take frogs, waterbird chicks, insects, and occasionally birds, lizards, and small mammals. They deftly snag prey from the water after a short flight and consume it back on the perch.

Active flight is with deep, strong wing beats. They soar often with wings held in a slight dihedral and glide with wings somewhat bowed. Display flights include undulating flight, often accompanied by vocalizations.

Usual vocalization is similar to that of Snail Kites, a grunting "aaaaahhhh."

MOLT Not studied. Apparently the annual molt of body, coverts, and tail feathers is complete, but not all primaries and secondaries are replaced annually, the former resulting in wave molt. Juveniles molt slowly into adult plumage, beginning at around nine months of age.

DESCRIPTION Sexes are alike in plumage, but females are somewhat larger. Eyes and cere are dark, and legs are gray. **Adult.** White head, black necklace, rufous body and underwing coverts, and dark brown remiges are diagnostic. Tail is banded rufous and black with a black subterminal band. Wide wings of adults make tail appear short. Rufous inner secondaries and undertail show dark barring. **Juvenile.** White head and dark collar are like adult. Narrower wings lack a strong rufous cast, and whitish underparts are marked brownish, usually with a white breast patch. Tail is similar to adults', but with less rufous.

FINE POINTS They have spiny bottoms on their toes and highly curved talons, similar to those of Ospreys, that help them to hold slippery fish.

UNUSUAL PLUMAGES No unusual plumages have been described.

HYBRIDS No hybrids have been described.

ETYMOLOGY *Busarellus* comes from the French *buse*, "buzzard," and *rayèe*, "streaked." The species name, *nigricollis*, comes from the Latin *nigri*, "black," and *collis*, "necked."

BELOW LEFT: Black-collared Hawk. Adult. White face, black collar, and rufous plumage are distinctive. Brazil. August 2012. WSC

BELOW RIGHT: Black-collared Hawk. Adult. White head and rufous plumage are distinctive. Brazil. August 2012. WSC

Black-collared Hawk. Juvenile. Juveniles are similar to adults but are much less rufous overall. Belize March 2011. RICHARD PAVEK

NORTHERN HARRIER PLATE 9
Circus hudsonius

VARI NORTEÑO

LENGTH 41–50 cm   **WINGSPAN** 97–122 cm   **WEIGHT** 290–600 g

IDENTIFICATION SUMMARY Slim-bodied raptor with long legs, long wings, and a long tail. Their distinctive quartering flight, flying low over the ground, is unique. White patch on the uppertail coverts and a dark head that appears hooded and shows an owl-like facial disk are distinctive. Sexes have different adult plumages, but are nearly identical in juvenile plumage. Females are noticeably larger than males. On perched harriers, wingtips do not reach tail tip.

TAXONOMY AND GEOGRAPHIC VARIATION Monotypic species, with no geographic variation. Formerly considered a subspecies of the Eurasian Hen Harrier, *C. cyaneus*, of Eurasia, but they differ in plumages sufficiently to warrant separate species. (See Oatley et al. 2015.)

SIMILAR SPECIES (1) **Turkey Vulture** (plate 1) also flies with wings in a dihedral, but is larger and overall blackish, has unbanded silvery undersides of flight feathers, and lacks white uppertail coverts.

(2) **White-tailed Kite** (plate 5) resembles adult male Northern Harrier in being black, white, and gray, but has small black carpal mark on underwings, black shoulder when perched, black patch on upperwings when in flight, and white tail, and lacks gray hood, black trailing edge of wing, and white uppertail coverts.

Northern Harrier. Adult male. Adult males are gray, white and black, with a variable amount of rufous markings on the underparts. California USA. January 2012. RICHARD PAVEK

(3) **Long-winged Harrier** (plate 32), a vagrant in Panama, is a similar-looking harrier; see under that species for distinctions.

STATUS AND DISTRIBUTION Passage migrant and winter visitor to all of our area. However, there is a small local breeding population in n Baja California. In Central America, they are more common on the Pacific slope.

HABITAT Occurs in a variety of open habitats, especially grassy ones, but also marshes. Migrants can fly over any habitat.

BEHAVIOR They hunt most of the time with their distinctive quartering flight, flying low over the ground and pouncing quickly when prey is spotted. However, they can fly directly toward avian prey in a rapid flight from some distance, with a short, twisting tail chase at the end. Males prey more on birds, while females take more mammals, but they both take both. They have been reported to drown waterfowl. Recent studies have shown that Northern Harriers can locate prey by sound almost as well as owls can, which explains the facial disk.

Active flight is with slow wing beats of flexible wings. Harriers soar usually with wings in a slight dihedral, but also on flat wings, and can appear somewhat buteo-like when soaring and gliding at high altitudes. Glides with wings in a modified dihedral, but hunts with wings in a strong dihedral. When soaring and quartering in strong winds,

Northern Harrier. Adult female. Adult females have brown upperparts, heavily streaked underparts, white bands on underwings, and yellow eyes. California USA. June 2011.
RICHARD PAVEK

141

harriers rock, somewhat like a Turkey Vulture. Courtship flights of males are spectacular, with steep dives and climbs and a series of rapid loops with the bird upside down at the top of each loop.

They are usually silent when not breeding, but can utter short calls when disturbed.

MOLT Annual molt is usually complete. Juveniles molt into adult plumage beginning after they migrate back to the breeding grounds.

DESCRIPTION Adult males and females have different plumages, and juvenile plumage is similar to adult females'. Cere is greenish yellow to yellow. Legs are orange-yellow. **Adult male.** Head is medium gray. Eyes are bright yellow. Back and upperwing coverts are darker gray. Gray of head extends onto upper breast so that head appears hooded. Rest of underparts are white with rufous spotting on breast, sometimes heavy and extending onto belly. Underwings are white, except for black tips of outer primaries and dark band formed by black tips of secondaries. Leg feathers and undertail coverts are white, often covered with small rufous spots. Long tail is medium gray above and whitish below and is indistinctly banded. **Adult female.** Head, back, and upperwing coverts are dark brown, with tawny mottling on head and upperwing coverts, especially tawny streaking on the neck. Eyes are brown to yellow (it takes from two to six years for them to become completely yellow). Underwings show heavily barred flight feathers and secondaries, and coverts form a dark patch on each underwing. Underparts are white to cream, completely marked with dark brown streaks, with some dark barring on the flanks and axillars. Leg feathers and undertail coverts are white with dark brown streaks and spots. Long tail is marked with even-width light and dark brown bands, with central pair of feathers noticeably darker. **Juvenile.** Similar to adult female, but appears darker, with less tawny mottling on head, neck, and upperwing coverts. Eye color of female is chocolate brown, whereas that of male is light gray or light gray brown. Underparts in fall are dark rufous, but fade nearly to white by spring and have dark streaking restricted to upper breast. Underwing pattern is like that of adult female, but with even darker secondary patches. Leg feathers and undertail coverts are unmarked rufous. Tail is like that of adult female.

FINE POINTS Juveniles' iris color is dimorphic: chocolate brown in females, gray to gray-brown in males.

UNUSUAL PLUMAGES Three melanistic harriers have been reported in w US; all were overall dark, sooty charcoal-gray, with whitish remiges below, and lacked white uppertail coverts. Photographs of an amelanistic harrier were taken in Rhode Island.

HYBRIDS No hybrids have been reported.

ETYMOLOGY Circus is from the Greek *kirkos*, "circle," from its habit of flying in circles. The species name, *hudsonius*, is after Hudson Bay, where the first specimen was collected. It was formerly called the Marsh Hawk. "Harrier" comes from the Old English *hergian*, "to harass by hostile attacks."

REFERENCES Oatley et al. 2015.

OPPOSITE PAGE: Northern Harrier. Juvenile male. Juveniles are similar to adult females but have rufous, usually unstreaked, underparts, gray bands on underwings, and gray or pale brown (males) or dark brown (females) eyes. California USA. December 2010.
RICHARD PAVEK

LONG-WINGED HARRIER PLATE 32

Circus buffoni

VARI ALAS LARGAS

LENGTH 43–60 cm   **WINGSPAN** 119–145 cm   **WEIGHT** 380–645 g

IDENTIFICATION SUMMARY Vagrant from South America, with two recent records of light-morph harriers in e Panama. All show pale facial ring, pale gray primary panels on upperwings, and white uppertail coverts. Only the light morph is depicted.

TAXONOMY AND GEOGRAPHIC VARIATION Monotypic, with no geographic variation.

SIMILAR SPECIES (1) **Northern Harrier** (plate 9) is similar in size, shape, and behavior. Adult males have gray head and upperparts and white underparts with rufous spotting. Adult females are overall brown, with heavily streaked underparts. Juveniles have dark brown head and upperparts and show sparse streaking on buffy to rufous underparts. They always lack pale gray primary panels on upperwings

Long-winged Harrier. Juvenile. Juveniles have dark upperparts and streaked underparts. Note white facial disk, gray on primaries, and wingtips reaching tail tip. French Guinea. January 2009.

MICHEL GIRAUD-AUDINE

144

STATUS AND DISTRIBUTION Vagrants from South America. Eight Panama records are from just east of Panama City to Darién near the Colombia border.

HABITAT Occurs in a variety of wet open habitats, including rice fields, marshes, wet grasslands, wet savannahs, and even crop lands.

BEHAVIOR They are typical harriers that hunt in low, slow flight just above the ground, pouncing on prey with a quick agile strike. They eat small mammals, small birds, frogs, lizards, and bird's eggs. Some prey is located by hearing, enhanced by their owl-like facial disk.

Active flight is with slow wing beats of flexible wings. They soar with wings in a slight dihedral, but also on flat wings, and can appear somewhat buteo-like when soaring and gliding at high altitudes. They glide with wings in a modified dihedral, but hunt with wings in a strong dihedral. When soaring and quartering in strong winds, harriers rock, somewhat like a Turkey Vulture.

MOLT Annual molt is usually complete. Juveniles molt into adult plumage beginning seven to nine months after fledging.

DESCRIPTION Typical harrier, with long wings and tail. They show an owl-like facial disk. They have two color morphs, but only the light morph has been recorded in Panama; the light morph only is described below. **Adult male.** Head and upperparts are black, and white underparts are unmarked. Underwings show white coverts marked with fine blackish barring. Remiges are gray above and white below, with a variable amount of rufous wash on base of primary remiges, and are narrowly banded black, with wide terminal bands on secondaries. Uppertail coverts are white. Long tail has black and gray bands of equal width above, but folded undertail is white with narrow black bands. Eyes are medium brown. **Adult female.** Similar to adult male, but head and upperparts are brownish-black with pale feather edges, and white underparts show a narrow streaking on upper breast and flanks. Eyes are pale to medium brown. **Juvenile, lightly marked.** Head and upperparts are brown, and pale buffy underparts show faint narrow streaking on upper breast and flanks. Remiges show gray primaries and brownish secondaries above and are pale gray below and narrowly banded black. Underwings show buffy coverts heavily marked with dark brown bars. White uppertail coverts are lightly marked. Long tail has dark brown and gray bands of equal width above, and dark brown and gray bands of equal width below. Eyes are dark brown. **Juvenile, heavily marked.** Some individuals show heavily streaked buffy underparts.

UNUSUAL PLUMAGES No unusual plumages have been described.

HYBRIDS No hybrids have been reported.

ETYMOLOGY *Circus* is from the Greek *kirkos*, "circle," from its habit of flying in circles. The species name, *buffoni*, is for French naturalist George Louis Leclerc, Comte de Buffon.

Long-winged Harrier. Juvenile. Upperwings are somewhat two-toned—paler primaries contrast with darker secondaries. Argentina. July 2011.
SERGIO SEIPKE

GRAY-BELLIED HAWK PLATE 32

Accipiter poliogaster

GAVILÁN VIENTRE GRIS

LENGTH 38–46 cm   **WINGSPAN** 69–84 cm   **WEIGHT** no data

IDENTIFICATION SUMMARY Medium-sized forest goshawks widespread in tropical South America. A juvenile occurred as a vagrant in Costa Rica. Adults are gray and white, and the juvenile mimics the striking plumage of adult Ornate Hawk-Eagles. Wingtips extend about one-third of the way down the tail on perched hawks.

TAXONOMY AND GEOGRAPHIC VARIATION Monotypic, with no geographic variation.

SIMILAR SPECIES (1) **Slaty-backed Forest-Falcon** (plate 28) adults are also gray and white like adult Gray-bellied Hawks, but they have paler gray head and upperparts, dark eyes, longer legs, and tail with graduated tip and pale bands that include dark lines.

(2) **Bicolored Hawk** (plate 11) juveniles are somewhat similar, but they have a pale hind collar, creamy underparts, buffy leg feathers, and wider pale tail bands.

(3) **Ornate Hawk-Eagle** (plate 25) adults are similar in plumage to juveniles, but they are much larger and have long crests and feathered tarsi.

STATUS AND DISTRIBUTION Vagrant to Costa Rica and Panama, with two records each. Fairly widespread in lowland forest in South America from Colombia to n Argentina.

HABITAT Occurs in lowland tropical forest and humid secondary forest.

BEHAVIOR They are little known. Based on limited sightings and foot structure, they seem not to be a specialist in catching birds and most likely take a wide variety of prey, including mammals, reptiles, insects, and amphibians, as well as birds.

Active flight is with a few deliberate wing beats, then a short glide. They have not been reported to soar.

MOLT Not studied. Presumably molt is completed annually. Juveniles molt directly into adult plumage.

DESCRIPTION A large, distinctive accipiter. Sexes are alike in plumage, but females are larger. Juvenile plumage is unique in that it mimics the plumage of adult Ornate Hawk-Eagle. **Adult.** Head is blackish, and cheeks can be black or gray. Upperparts are dark gray, and underparts are white. Underwings show unmarked white coverts and narrowly banded remiges. Black tail shows three narrow gray bands. Long legs are yellow. Eyes are yellow. Cere, gape, and face skin are orange-yellow. **Juvenile.** Striking. Plumage mimics adult Ornate Hawk-Eagle. Head shows black crown (but not long crest);

rufous cheeks and nape, the latter extending onto sides of the breast; narrow black malar stripes; and a white throat. White belly and upper leg feathers are boldly barred black and white, and white undertail coverts have black tips. White underwing coverts show some small black spots, and whitish remiges are narrowly banded below. Upperparts are blackish, with white feather edges, becoming more grayish on uppersides of remiges. Long tail above shows blackish and gray-brown bands of equal width and below appears whitish with narrow black bands. Long legs are bare and dull yellow. Eyes are dull yellow, and cere is greenish.

UNUSUAL PLUMAGES No unusual plumages have been described.

HYBRIDS No hybrids have been reported.

ETYMOLOGY *Accipiter* is Latin for "bird of prey." It probably derives from *accipere*, "to take," but also possibly from the Greek *aci*, "swift," and *petrum*, "wing." The species name, *poliogaster*, is from the Greek *polios*, "gray,'" and *gastros*, "belly."

ABOVE: Gray-bellied Hawk. Juvenile. Juveniles look remarkably like adult Ornate Hawk-Eagles, but note bare tarsi. Brazil. October 2013. KEVIN ZIMMER

LEFT: Gray-bellied Hawk. Adult. Adults have gray heads and upperparts and unmarked white underparts. Costa Rica. April 2014. CHRISTIAN CONTRERAS CASTRO

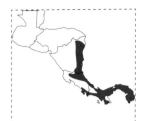

TINY HAWK PLATE 10

Hieraspiza superciliosa

GAVILÁN CHICO

LENGTH 20–28 cm   **WINGSPAN** 38–48 cm   **WEIGHT** 75–135 g

IDENTIFICATION SUMMARY Small, thrush-sized hawks, with noticeable yellow supra-orbital ridges. Adult plumage is gray above and barred gray below (sexes alike). Juvenile plumage is dimorphic: brown morph is common, and rufous morph is rare. Both have barred underparts. Females are much larger than males. They do not soar, and their flight is direct on rapid wing beats. Primary projection is short, barely extending beyond the folded secondaries. Short tail shows square corners on tips. Their legs are relatively shorter than those of small accipiters.

TAXONOMY AND GEOGRAPHIC VARIATION *H. s. fontanieri* occurs from Nicaragua to Panama. Formerly included in the genus *Accipiter*, but it differs considerably from hawks in that genus in bone structure (see Olson 2006).

SIMILAR SPECIES (1) **Sharp-shinned Hawk** (plate 10) is a similar hawk, but it is larger and has longer legs and tail. Adults have rufous underparts, and juveniles have streaked underparts.

(2) **Double-toothed Kite** (plate 6) adults are similar, but are larger and show a dark throat stripe, and they have longer broader wings with a heavy, blackish tail with three narrow whitish bands.

Tiny Hawk. Adult male. Adults have blue-gray crown and upperparts, and gray barred underparts. Males are much smaller than females. French Guinea. January 2014.

MICHEL GIRAUD-AUDINE

(3) **Barred Forest-Falcon** (plate 28) is similar; see under that species for distinctions.

STATUS AND DISTRIBUTION Uncommon to rare and unobtrusive (and overlooked) in the middle to upper stories of all but densest tropical forests from e Nicaragua south to Panama. They are more common on the Caribbean slope north and west of the Panama Canal.

HABITAT Occurs in lowland tropical forest, second growth, and plantations, and prefers mid to upper levels, especially on forest edge and other open areas.

BEHAVIOR They use concealed perches to hunt for their prey of birds (occasionally small mammals), stooping in rapid flight to capture them. They are reported to capture hummingbirds.

Active flight is direct with a series of rapid wing beats, interspersed with short glides. They do not soar.

A few stick nests in tall trees have been reported.

MOLT Not studied. Annual molt is apparently complete. Juveniles begin a slow molt into adult plumage about a year after fledging.

DESCRIPTION Plumage is the same by sex, but females are considerably larger. Cere and lores are yellow. Legs are yellow-orange. Short tail shows square corners on tips. **Adult.** Gray head shows darker crown, red-orange eyes, and bright yellow cere and lores. Upperparts are dark blue-gray, and white underparts are narrowly barred dark gray. Short tail shows black and gray bands of equal width on uppersides. Whitish undersides of folded tail (outer feathers) are unbanded. **Juvenile brown morph** (common). Brown head has darker crown. Upperparts are dark brown, and creamy underparts show narrow rufous barring, heavier on leg feathers. Short tail shows dark brown and brown bands of equal width. Juveniles have yellow to orange eyes. **Juvenile rufous morph** (rare). Rufous head has dark crown. Rufous upperparts show some brown markings, and creamy underparts show narrow rufous barring, heavier on leg feathers. Rufous tail has narrow dark brown bands.

FINE POINTS Legs are occasionally yellow.

UNUSUAL PLUMAGES No unusual plumages have been described

HYBRIDS No hybrids have been described.

ETYMOLOGY *Hieraspiza* comes from the Greek *hierax*, "hawk," and *spizias*, also "hawk." The species name, *superciliosa*, comes from the Latin *superciliosus*, "eye-browed," in reference to its yellow supra-orbital ridges. Formerly considered to be in the genus *Accipiter*, but it differs in bone structure.

REFERENCES Olson 2006.

Tiny Hawk. Juvenile rufous. Juveniles have brown or rufous crown and upperparts and brown or rufous barred underparts. Rufous juveniles are much less common. Costa Rica. July 2007. CHRIS WOOD

SHARP-SHINNED HAWK PLATE 10

Accipiter striatus

GAVILÁN NORTEÑO

LENGTH 24–35 cm   **WINGSPAN** 52–68 cm   **WEIGHT** 87–218 g

IDENTIFICATION SUMMARY Small accipiters that are breeding residents of pine forest in mountain forests of c and s Mexico and widespread throughout our area on migration and during winter. Sexes are almost alike in plumage, but females are separably larger than males. Head does not project much beyond the wrists of wings that are thrust forward when they are in flight. Tail feathers are about the same length, so tip appears straight across, with squared corners, especially on males, and tip shows a narrow white terminal band and often appears notched. Cere is yellow green to yellow. Long, stick-like legs are yellow. Head almost always appears rounded as hackles are never raised, and wingtips reach less than halfway to tail tip on perched hawks.

TAXONOMY AND GEOGRAPHIC VARIATION Two subspecies occur: *A. s. velox* is widespread on migration and in winter, and *A. s. suttoni* is resident in c and s Mexico. They differ little.

SIMILAR SPECIES (1) **Cooper's Hawk** (plate 11), in spite of size and structure differences, appears very similar to Sharp-shinned, but is more robust; has thicker legs; has longer, more rounded tail with a wide white terminal band; and often raises hackles on rear crown. Head projects far beyond wrists on gliding birds, and they often soar with leading edge of wings straight out from the body and with wings in a dihedral. Adult's crown is darker than nape and back, with a noticeable line of contrast. Juveniles' underparts are more finely streaked, with fewer belly markings; their backs have more whitish mottling, and cheeks may be tawny (always brown on immature Sharp-shinned).

(2) **White-breasted Hawk** (plate 10) is similar in size and behavior; see under that species for distinctions.

(3) **Double-toothed Kite** (plate 6) is similar in size and flight shape; see under that species for distinctions.

STATUS AND DISTRIBUTION Uncommon to fairly common breeding residents in mountains of c and s Mexico north and west of the Isthmus of Tehuantepec, and widespread and fairly common on migration and in winter throughout most of our area.

HABITAT Breeds in pine-oak forests in Mexico, and occurs in many habitats elsewhere.

BEHAVIOR These shy and retiring hawks hunt from inconspicuous perches in wooded areas for small birds, almost their only prey, which they capture after a brief, rapid chase. They also hunt by coursing

Sharp-shinned Hawk. Juvenile. Juveniles have brown crown and upperparts and rufous to brown streaked underparts. Males are much smaller than females. Texas USA. November 2008.
JIMMY PAZ

over or through the woods, hoping to surprise their victims. On migration, they move by active flight until thermals form and thereafter soar, often up to an altitude of thousands of feet.

They are unable to raise their hackles, and so their heads appear rounded.

Active flight is light and buoyant, with rapid, soft, deep wing beats. They soar and glide on level wings, but with wrists pushed forward. They are extremely agile in pursuing small birds.

They build small stick nests, most often in smaller evergreens.

MOLT Annual molt of adults is complete. Juveniles molt directly into adult plumage during their second summer. Usually they retain a few juvenile upperwing or uppertail coverts in the first adult plumage.

DESCRIPTION The sexes are alike in plumage. **Adult (A. s. velox).** Crown, nape, back, and upperwing coverts are blue-gray, brighter on males and more pronounced browner on females by spring and summer. Crown is the same color or slightly darker than blue-gray back, lacking line of contrast with nape. Cheeks are rufous. Skin of supra-orbital ridge, if visible, is yellow. Eyes are orange to red. White underwing coverts and underparts, including leg feathers, are finely barred rufous. Pale undersides of flight feathers show bold dark barring. Tail has narrow white band on tip in fresh plumage, and upperside shows equal-width bands of dark and light brown, except for the outer pair, which have narrower dark bands. Undertail coverts are white. **Adult (A. s. suttoni).** Plumage is like *velox* except for uniformly rufous leg feathers. Underparts tend to be more heavily rufous, with less distinct barring. **Juvenile (subspecies alike).** Dark brown head shows long narrow pale superciliary lines and pale throat; cheeks are somewhat paler on some individuals, usually those with narrow dark streaks on underparts. Eyes are yellow, becoming orangish by spring. Back and upperwing coverts are dark brown with narrow rufous feather edging, usually with little or no white mottling. Whitish to creamy underparts of most females have thick rufous streaking, becoming thicker and more barred on flanks and belly, but streaking is a narrower dark brown to rufous-brown on most males and some females. Undersides of flight feathers show bold dark barring; some birds, more commonly females, have dusky rather than dark tips of flight feathers. Upperside of tail has equal-width bands of dark brown and paler gray-brown, except for outer feathers, which have narrower dark bands; all show a narrow pale terminal band in fresh plumage. Undertail coverts are white.

FINE POINTS "Sharp-shinned" refers to the raised ridge on the inside front of the tarsus (not actually a shin).

UNUSUAL PLUMAGES A partial albino with white wings has been reported, and amelanistic specimens have creamy or café-au-lait color instead of dark brown upperparts and only rufous markings on underparts.

HYBRIDS Specimens taken in s Mexico where the breeding ranges of Sharp-shinned and White-breasted Hawks meet are intermediate.

ETYMOLOGY *Accipiter* is Latin for "bird of prey." It probably derives from *accipere*, "to take," but also possibly from the Greek *aci*, "swift," and *petrum*, "wing." The species name, *striatus*, is Latin for "striped," a reference to the underparts of the juvenile, which was described prior to the adult. "Sharp-shinned" is apparently after the inner edge of the row of scales on the front of the tarsi, which form a noticeable ridge.

Sharp-shinned Hawk. Juvenile. Juveniles have rufous to brown streaked underparts that extend onto the belly. Texas USA. September 2007. WSC

WHITE-BREASTED HAWK PLATE 10
Accipiter chionogaster
GAVILÁN VIENTRE BLANCO

LENGTH 24–35 cm **WINGSPAN** 52–68 cm **WEIGHT** 90–218 g

IDENTIFICATION SUMMARY Small accipiters that are resident breeders in mountain pine forests from se Mexico to nw Nicaragua. Their whitish underparts are unmarked or faintly marked. Sexes are almost alike in plumage, but females are separably larger than males. Long tail has square corners on tips. Head almost always appears rounded as hackles are never raised, and wingtips reach less than halfway to tail tip on perched hawks. Cere is yellow, and long, stick-like legs are orangish-yellow, brighter on adults.

TAXONOMY AND GEOGRAPHIC VARIATION Monotypic, with no geographic variation. Formerly considered a subspecies of Sharp-shinned Hawk (*Accipiter striatus*), but with little justification, as they differ in plumages and habitat.

SIMILAR SPECIES (1) **Bicolored Hawk** (plate 11) juveniles are very similar in plumage to adults, but they are larger, have a wide whitish hind collar, yellow eyes, a rufous wash on leg feathers, and whitish pale bands on uppertail.

(2) **Sharp-shinned Hawks** (plate 10) are similar in shape and behavior, but have heavily marked underparts.

STATUS AND DISTRIBUTION Fairly common breeding residents in highland pine forests from se Mexico to nw Nicaragua. There are no records of altitudinal migrations, but some have dispersed some distances.

HABITAT Occurs almost exclusively in highland pine forest, especially *Pinus oocarpa*, but also in higher pine forest. They are not found in lowland pine forest. They hunt in adjacent cloud forest, tropical dry forest, and farmlands.

BEHAVIOR These typical bird-hunting accipiters hunt from inconspicuous perches in wooded areas looking for small birds, almost their only prey, which they capture after a brief, rapid chase. They also hunt by coursing over or through the woods, hoping to surprise their victims. They also take bats, lizards, grasshoppers, and insects, especially when breeding.

They almost never raise their hackles, so their heads appear rounded.

Active flight is light and buoyant, with rapid, soft, deep wing beats. They soar and glide on level wings, but with wrists pushed forward.

White-breasted Hawk. Adult. Adults have dark blue-gray crown and upperparts and unmarked white underparts. Males are much smaller than females. Guatemala. December 2014.
KNUT EISERMANN

154

They build a small stick nest in pine trees. Jenner (2010) describes their breeding in detail.

MOLT Annual molt of adults is complete, but is suspended while breeding. Juveniles undergo an early molt into adult plumage, completed before the next breeding season.

DESCRIPTION Sexes are alike in plumage. Cere is yellow, and long, stick-like legs are orangish-yellow, brighter on adults. **Adult.** Dark on crown extends onto sides of head, enclosing the eyes, but not extending into the white auriculars. Dark blue-gray upperparts contrast with white throat, body, and underwing and undertail coverts. Whitish remiges have bold, narrow dark barring. Wings are short and broad. Uppertail shows black and gray bands of equal width. Undertail shows black and white bands of equal width, with narrow black bands on outer feathers. Eyes vary from orange (young adults) to red to dark brownish-red. **Juvenile female.** Dark brown upperparts contrast with whitish underparts that are faintly streaked rufous. Leg feathers are noticeably rufous. Uppertail shows dark brown and grayish-brown bands of equal width. Folded undertail shows narrow black bands. Eyes begin as pale gray then turn to dull yellow and brighten with age.

FINE POINTS One-year-olds are in complete adult plumage, but have orange or pale red eyes.

UNUSUAL PLUMAGES No unusual plumages have been described.

HYBRIDS Specimens taken in s Mexico where the breeding ranges of Sharp-shinned and White-breasted Hawks meet are intermediate.

ETYMOLOGY *Accipiter* is Latin for "bird of prey." It probably derives from *accipere*, "to take," but also possibly from the Greek *aci*, "swift," and *petrum*, "wing." The species name, *chionogaster*, comes from the Greek *khion*, "snow," and *gaster*, "belly," for its white underparts.

REFERENCES Jenner 2010.

BELOW LEFT: White-breasted Hawk. Adult. Adults have dark blue-gray crown and unmarked white underparts. Guatelama. December 2014.
KNUT EISERMANN

BELOW RIGHT: White-breasted Hawk. Juvenile. Juveniles are like adults but with faint breast streaking and brown crown and upperparts (not shown). El Salvador. June 2008.
WSC

COOPER'S HAWK PLATE 11

Accipiter cooperii

AZOR DE COOPER

LENGTH 37–47 cm **WINGSPAN** 70–87 cm **WEIGHT** 302–678 g

IDENTIFICATION SUMMARY Robust, medium-sized accipiters. Sexes are almost alike in plumage, with females separably larger than males. Adult and juvenile plumages differ. On gliding birds, head projects far beyond wrists. They usually soar with the leading edges of the wings held perpendicular to the body and forming a straight line, and with wings held in a slight to medium dihedral. Outer tail feathers are progressively shorter than the central pair. Corners of tail tip appear rounded, and tip shows a broad white terminal band. On perched birds, head often appears squarish due to raised hackles, but head appears rounded when they are not raised, and wingtips reach less than halfway to tail tip.

TAXONOMY AND GEOGRAPHIC VARIATION Monotypic, with no geographic variation.

SIMILAR SPECIES (1) **Sharp-shinned Hawk** (plate 10), in spite of size difference, appears very similar in all plumages to Cooper's Hawk; see under that species for distinctions.

(2) **Northern Goshawk** (plate 11) juvenile is similar, but is larger and has relatively longer and more tapered wings, more heavily streaked underparts, wide pale superciliary lines, markings on undertail coverts, a paler, more mottled back, and irregular, wavy tail bands. Adult goshawks have a pale blue-gray breast.

(3) **Gray Hawk** (plate 17) juveniles are similar to juveniles in flight; see under that species for distinctions.

(4) **Broad-winged Hawk** (plate 18) adults and juvenile are similarly colored, but have dark malar stripes, pointed wings lacking strong banding on undersides of remiges, and a shorter tail with different patterns.

(5) **Red-shouldered Hawk** (plate 17) juveniles are accipiter-like, but have buteo-shaped wings with crescent-shaped primary panels, dark malar stripes, and tail with dark and light bands of unequal width.

STATUS AND DISTRIBUTION Rare and local resident breeders in the mountains of w and s Mexico (Sonora and Chihuahua south to Guerrera and Oaxaca). They are more common and widespread on migration and in winter south to Honduras, and are rare to vagrant farther south.

HABITAT Breeds in highland forests of pine and oak in Mexico. They migrate over all habitats and winter in a variety of open forests. Compared to other accipiters, they occur in more open habitats.

BEHAVIOR Primarily still hunters, they perch inconspicuously in woods or in the open on snags or poles, waiting to attack their prey at an opportune moment. However, they also hunt by flying through or over woodlands or along fencerows to surprise potential prey and pursue prey into dense cover, sometimes on foot. The preferred prey of Cooper's Hawk is birds, but many small mammals and lizards are also taken.

When perched, the birds often raise their hackles, so that the head appears larger and squarish.

Active flight is with stiff, strong, rather slow wing beats. They glide with wings held level, wrists cocked a bit forward, and soar with wings held in a slight to medium dihedral, with the leading edge of the wing straighter and more perpendicular to the body than those of other accipiters. They will soar for a short period on sunny days; migration is mostly by soaring and gliding.

They build a substantial stick nest in trees, recently in urban areas.

They vocalize usually only around the nest, emitting a series of loud "kak" or "kek" calls.

MOLT Annual molt of adults is complete. Juveniles molt directly into adult plumage in their second summer. Usually they retain a few juvenile upperwing or uppertail coverts in first adult plumage.

DESCRIPTION Sexes are almost alike in plumage, with females separably larger than males. Cere is yellow green to yellow. Stout

BELOW LEFT: Cooper's Hawk. Adult. Adults have dark crowns and paler napes. Cheeks are gray on older males but rufous on females and younger males. California USA. March 2005. JIM GALLAGHER—SEA & SAGE AUDUBON

BELOW RIGHT: Cooper's Hawk. Juvenile. Juveniles usually do not show a pale superciliary line. Streaking is narrower and sparser on their bellies. California USA. June 1997. WSC

Cooper's Hawk. Juvenile. Streaking on underparts is sparser on juveniles' bellies. Note the straight leading edge of wings during glides. Arizona USA. September 2010.

NED HARRIS

legs are yellow. **Adult.** Crown is blackish and contrasts with pale nape and blue-gray back. Eyes are orange to red. Cheeks are rufous on females and first-plumage adult males, but are gray on older males. Skin of supra-orbital ridge, if visible, is grayish. Back and upperwing coverts are blue-gray on males and gray on females, latter becoming more brownish by spring and summer. White underwing coverts and underparts, including leg feathers, are barred rufous. Flight feathers are heavily barred below, but band on tip is usually dusky, paler than the other dark bands. Many males show muted barring on secondaries. Upperside of long tail has equal-width bands of blackish-brown and gray-brown, except for the outer pair, which have more and narrower dark bands; all show a broad white terminal band in fresh plumage. Undertail coverts are white. **Juvenile.** Brown head usually lacks pale superciliary lines; many individuals have a tawny wash on cheeks. Back and upperwing coverts are dark brown, but appear paler due to whitish mottling and wide rufous feather edging. Underparts, underwing coverts, and leg feathers are white with narrow dark brown streaking, becoming sparse or absent on the belly. Flight feathers are heavily barred below. Upperside of long tail has equal-width bands of dark brown and paler gray-brown, except for the outer pair, which have narrower dark bands; all show a broad white terminal band in fresh plumage. White undertail coverts are usually unmarked, but a few western hawks show narrow, dark shaft streaks here. Eyes are light greenish-gray to dull greenish-yellow, becoming a brighter yellow by spring.

FINE POINTS Cooper's Hawk's long tarsus is feathered only a third of the way to the feet, but Northern Goshawk's is feathered halfway to the feet.

UNUSUAL PLUMAGES Amelanistic plumage specimens have creamy or café-au-lait color instead of dark brown upperparts and only rufous markings on underparts. Two instances of partial melanistic Cooper's Hawks have been reported in the US (Schroeder 2009 and Morrow et al. 2015).

HYBRIDS Several adult hybrids between Northern Goshawks and Cooper's Hawks have been captured for banding (e.g., Gunther and Hopey 2010).

ETYMOLOGY *Accipiter* is Latin for "bird of prey." It probably derives from *accipere*, "to take," but also possibly from the Greek *aci*, "swift," and *petrum*, "wing." "Cooper's" and *cooperii* are after New York ornithologist William Cooper.

REFERENCES Gunther and Hopey 2010, Morrow et al. 2015, Schroeder 2009.

BICOLORED HAWK PLATE 11

Accipiter bicolor

AZOR BICOLOR

LENGTH 34–45 cm **WINGSPAN** 61–79 cm **WEIGHT** 200–460 g

IDENTIFICATION SUMMARY Medium-sized, inconspicuous forest accipiter. Sexes are alike in plumage, but females are separably larger. Adults are overall gray, darker above and paler below, with rufous leg feathers. Juveniles are dark brown above and creamy below, with noticeable white hind collar; both lack markings on underparts. Long blackish tail shows three narrow white bands. They do not soar. Wingtips barely extend to halfway down long tail on perched hawks.
TAXONOMY AND GEOGRAPHIC VARIATION *A. b. bicolor* occurs in our area, with no geographic variation. The validity of *A. b. fidens* is questionable.

Bicolored Hawk. Adult. Adults have dark blue-gray crown and upperparts, paler blue-gray underparts, and rufous leg feathers. Costa Rica. December 2012. CHRIS JIMÉNEZ

SIMILAR SPECIES (1) **Collared Forest-Falcon** (plate 28) adults are similar to perched juveniles in coloration and having a pale hind collar, but are larger, have dark eyes and cere, and a slash across the cheeks, and their longer tail is graduated. Their primary projection is barely beyond the folded secondaries.

(2) **White-breasted Hawk** (plate 10) adults appear similar to perched juveniles; see under that species for distinctions.

(3) **Slaty-backed Forest-Falcon** (plate 28) adults appear similar to perched juveniles; see under that species for distinctions.

STATUS AND DISTRIBUTION Uncommon resident in lowland forests, seeming to be as rare as they are unobtrusive, from sw Tamaulipas, Mexico, to Panama on the Caribbean slope and on the Pacific slope of Costa Rica and Panama.

HABITAT Occurs in a variety of lowland and foothill tropical forests.

BEHAVIOR These still hunters specialize in preying on birds, but they also take small mammals, lizards, and large insects. They also fly into fruiting trees full of birds to scatter them and grab one.

Active flight is typical of accipiters: direct with three to five wing beats, then a short glide. They have not been recorded soaring.

They vocalize usually only around the nest, emitting a series of loud "kak" or "kek" calls.

MOLT Not studied. Annual molt is apparently complete. Juveniles begin molting slowly into adult plumage when less than a year old.

DESCRIPTION Sexes are alike in plumage, but females are separably larger. Long legs are yellow. **Adult.** Head and upperparts are dark gray, with blackish crown. Underparts are a medium gray, except for rufous legs and white undertail coverts. Long blackish tail shows three narrow bands: gray above and white below. Eyes are orange to red. Cere is yellow. **Buffy juvenile** (common). Dark brown head has darker crown and buffy throat. Dark brown upperparts have a wide buffy collar across hind neck and pale feather edges. Buffy underparts are unmarked, but leg feathers show a pale rufous-buff wash. Long blackish tail shows three narrow white bands. Eyes are yellow. Cere is dull yellow. **Rufous juvenile** (rare). Like buffy juvenile, but underparts and underwing coverts are rufous.

UNUSUAL PLUMAGES No unusual plumages have been described.

HYBRIDS No hybrids have been described.

ETYMOLOGY *Accipiter* is Latin for "bird of prey." It probably derives from *accipere*, "to take," but also possibly from the Greek *aci*, "swift," and *petrum*, "wing." The species name, *bicolor*, comes from the Latin *bi*, "two," and *color*, "color," in reference to the rufous leg feathers and gray underparts of adults.

OPPSITE PAGE: Bicolored Hawk. Juvenile. Juveniles have dark brownish-gray crown and upperparts and unmarked buffy underparts. All show narrow white tail bands. Costa Rica. September 2013. CHRIS JIMÉNEZ

NORTHERN GOSHAWK PLATE 11

Accipiter gentilis

AZOR NORTEÑO

LENGTH 46–62 cm **WINGSPAN** 98–104 cm **WEIGHT** 615–1,400 g

IDENTIFICATION SUMMARY Largest accipiter in our area and a breeding resident in Mexican western mountain forests. Adult and juvenile plumages differ. Sexes are almost alike in plumage, with females separably larger than males. All show wide bold superciliary lines. Wings are relatively long for an accipiter, rather more buteolike. Tip of folded tail is wedge-shaped. On perched birds, wingtips extend halfway to tail tip.

TAXONOMY AND GEOGRAPHIC VARIATION *A. g. atricapillus* occurs in Mexico, with no variation.

SIMILAR SPECIES (1) **Cooper's Hawk** (plate 11) juvenile is similar to juvenile Northern Goshawk; see under that species for distinctions.

(2) **Red-shouldered Hawk** (plate 17) juvenile usually has pale superciliary lines like juvenile Northern Goshawk, but has pale, crescent-shaped primary panels, dark malar stripe, and dark and light tail bands of unequal width.

(3) **Buteos** in flight are somewhat similar, but have a shorter tail and lack tapered wings. On perched buzzards, wingtips extend to or almost to tail tip.

STATUS AND DISTRIBUTION Rare and local resident in the Sierra Madre Occidental mountains of Mexico from Sonora and Chihuahua south to Jalisco and locally in Guerrero. They are apparently absent from suitable habitat in the Sierra Madre Oriental mountains of e Mexico, but a recent record there is valid, suggesting a small population.

HABITAT Occurs in conifer and pine-oak mountain forest.

BEHAVIOR They are found almost always in forest, where they are usually quite inconspicuous. They prey on medium to large mammals and birds and hunt from a perch or while flying through the forest. They also pursue prey on the ground. Goshawks courageously defend the nest area and will even strike humans, usually on the head or back. Like most accipiters, they soar for a while on most sunny days, usually in the morning.

Active flight is strong, with powerful, stiff wing beats; wingtips often appear rather pointed in active flight. They soar on level wings, appearing much like a buteo, and glide with wings level and wrists cocked somewhat forward.

They build large stick nests in tall trees.

They vocalize usually only around the nest, emitting a series of loud "kaah" calls.

162

MOLT Annual molt of adults is not complete; not all flight and tail feathers are changed every year. Adult females begin molt prior to egg laying, but usually suspend molt while breeding and complete it after the young are independent. Males wait until the young are well on the wing to begin molt. Juveniles molt into Basic II (second) plumage in their second summer, beginning later than adults. Hawks in Basic II plumage usually retain a few juvenile outer primaries, middle secondaries, or a tail feather or two, as well as some juvenile upperwing or uppertail coverts.

DESCRIPTION Adult and juvenile plumages differ. Sexes are almost alike in plumage, with females separably larger than males. Cere is greenish-yellow to yellow. Legs are yellow. **Adult.** Head consists of black crown, nape, and cheeks, a wide white superciliary line, and whitish throat. Eye color varies from orange to red to mahogany, darkening with age. Back and upperwing coverts are blue-gray and average darker on females; they contrast with the blackish uppersides of the flight feathers. Underparts are pale blue-gray, with fine black vermiculations and some vertical black streaking. Females usually have coarser, darker barring and more vertical black streaking. Primaries show dusky banding on undersides; secondaries show at most faint banding. Tail is blue-gray, with three or four incomplete blackish bands. Undertail coverts are white and fluffy. **Basic II (second plumage).** Almost like that of adult, but upperparts are darker, and they usually show heavier dark streaking and bolder barring on the underparts. Eyes are yellow-orange to orange-red. Undersides of flight feathers show distinct dark banding. Some brown juvenile outer primaries (seldom), middle secondaries (often), or tail feathers (sometimes) can be retained, as well as a few uppertail or upperwing coverts (usually). **Juvenile.** Wide buffy superciliary lines are noticeable on brown head. Eye color varies from pale green-yellow to yellow to orangish-yellow, rarely light brown. Brown back and upperwing coverts have extensive tawny and white mottling (and appear paler than those of Cooper's Hawks); a row of buffy spots is usually noticeable on the wing coverts of perched hawks and on the upperwing coverts of flying hawks. Buffy underparts are marked with wide blackish-brown streaks. White underwing coverts are heavily spotted and barred brown. Undertail coverts are marked with wide dark blobs (a few individuals have unmarked undertail coverts). Undersides of flight feathers are boldly banded with dark brown. Tail has irregular, wavy, dark and light brown bands of equal width,

Northern Goshawk. Adult female. Adults show distinctive head pattern of dark crown, bold pale superciliary line, and dark eye-line and cheeks. Underparts are barred, coarser on females and narrower on males. British Columbia, Canada. January 2009. WSC

with thin whitish highlights at many of the boundaries of these bands, and a wide pale terminal band. Dark bands on the outer pair of feathers are narrower than light ones. Underside of folded tail shows a zigzag pattern (however, some Cooper's and Sharp-shinned Hawks' undertails appear similar).

FINE POINTS Juvenile Northern Goshawks usually show a hint of a facial disk and sometimes a second shorter buffy wing bar. Goshawk's short tarsus is feathered halfway to feet (only one-third of the way in Cooper's Hawk).

UNUSUAL PLUMAGES Partial albinism, with some white feathers, has been reported, and amelanistic specimens have creamy or café-au-lait color instead of dark brown upperparts and only rufous markings on underparts.

HYBRIDS Several adult hybrids between Northern Goshawks and Cooper's Hawks have been captured for banding (Gunther and Hopey 2010).

ETYMOLOGY *Accipiter* is Latin for "bird of prey." It probably derives from *accipere*, "to take," but also possibly from the Greek *aci*, "swift," and *pteron*, "wing." The species name, *gentilis*, is Latin for "noble." It was named during the era when only the nobility could fly this bird in falconry. "Goshawk" was derived from the Anglo-Saxon words *gos*, "goose," and *havoc*, "hawk"—hence, a hawk that captures geese.

REFERENCES Gunther and Hopey 2010.

Northern Goshawk. Juvenile. Juveniles are heavily streaked below, including the undertail coverts. Note the narrowing of the wings at the primary-secondary junction. Utah USA. October 2007. JERRY LIGUORI

CRANE HAWK PLATE 9
Geranospiza caerulescens
GAVILÁN PATAS LARGAS

LENGTH 42–54 cm **WINGSPAN** 82–110 cm **WEIGHT** 230–430 g

IDENTIFICATION SUMMARY Distinctive, slender, long-legged dark raptor. They often reach into cavities with their long, double-jointed legs. Their paddle-shaped wings show a unique row of white spots forming a crescent on the wingtips. The adult plumage is the same for the sexes, and the juvenile plumage is similar. Their small heads sometimes show a short cowl. They occasionally soar. Wingtips barely extend beyond folded secondaries on perched hawks.

TAXONOMY AND GEOGRAPHIC VARIATION Two subspecies: *G. c. belzarensis* occurs in e Panama, and *G. c. nigra* occurs from w Panama north to s Mexico. Their plumages are similar, with the former being overall dark gray and the latter overall blackish and somewhat larger.

SIMILAR SPECIES (1) **Common Black Hawk** (plate 15) adults are also overall blackish with a white tail band, but are larger and stockier, show a small white flash at base of outer primaries, and lack row of white spots on outer primaries.

(2) **Zone-tailed Hawk** (plate 22) is also overall blackish with a white tail band, but it is larger, shows two-toned underwings, and lacks the row of white spots on the outer primaries.

(3) **Snail Kite** (plate 7) adult males are also overall blackish, but have orange cere and face skins, and an undertail that shows a white base rather than a band; they lack the row of white spots on the outer primaries.

(4) **Slender-billed Kite** (plate 7) is also overall dark gray like Crane Hawks in Panama; see under that species for distinctions.

(5) **Hook-billed Kite** (plate 4) dark-morph adults are also overall blackish with a white tail band, but show pinched wings and lack the row of white spots on the outer primaries.

STATUS AND DISTRIBUTION Uncommon and local breeding residents along both coasts of c Mexico and all of the Yucatán Peninsula south through Central America. They are absent in higher mountains.

HABITAT Occurs in a variety of forest habitats by or near water, including forest edge, gallery forest, marsh edge, and plantations.

BEHAVIOR They usually hunt from a perch and move on foot with agile grace among the branches of trees, but also hunt on the wing with a harrier-like flight over open areas,

Crane Hawk. Adult. Adults are overall gray, with red eyes and orange legs. Mexico. December 2009.
ANTONIO HIDALGO

especially marshes. They take a variety of prey, including tree frogs, lizards, small snakes, nestling birds, bats, rodents, spiders, and insects. They often reach into cavities, crevices, bromeliads, and long pendulous bird nests with their long, double-jointed legs, searching for prey.

Active flight is with slow, floppy wing beats. They soar and glide with wings level or in a slight dihedral.

They build their shallow and small stick nest in a tall tree, usually near the main trunk.

Usual vocalizations include a low, hollow "waah-o" given intermittently at dawn and dusk.

MOLT Not well studied. Annual molts are apparently complete. Juveniles begin their slow molt into adult plumage seven to nine months after fledging.

DESCRIPTION Sexes are alike in plumage, with females larger. Juvenile plumage is similar. Long legs are double-jointed. Bill and cere are dark. **Adults (*nigra*).** Head, body, and coverts are dark gray, occasionally showing fine white barring on belly and leg feathers, and occasionally a few small white spots on undersides. Outer primaries show a distinctive curve of white spotting near tips. Long black tail has two narrow to wide white bands. Eyes are carmine, and legs are reddish-orange to orange. **Juveniles (*nigra*).** Similar to adults, but with white face pattern and whitish barring on underparts. Back color is slightly more blackish. Eyes are orange, and legs are yellow-orange to orange. **Adults (*belzarensis*).** Almost like adult *nigra*, but overall a paler gray. Many show fine white barring on belly and leg feathers. **Juveniles (*belzarensis*).** Almost like juvenile *nigra*, but overall a paler gray.

FINE POINTS Cowl when raised gives small head a "bull-neck" appearance.

UNUSUAL PLUMAGES No unusual plumages have been reported.

HYBRIDS No hybrids have been reported.

ETYMOLOGY *Geranospiza* comes from the Greek *geranos* and *spizias*, "crane" and "hawk." The species name, *caerulescens*, comes from the Greek *caeruleus*, "blue." This species in South America is blue, not blackish as in our area.

BELOW: Crane Hawk. Adult. Adult using 'double-jointed' legs to probe into a cavity. Mexico. June 2008.
JIM ZIPP

BOTTOM: Crane Hawk. Juvenile. Juveniles are like adults, but show some white on the face and pale gray barring on the underwing coverts. All show row of white spots on the primaries. Mexico. March 2009.
JIM ZIPP

PLUMBEOUS HAWK PLATE 12
Cryptoleucopteryx plumbea

GAVILÁN PLOMIZO

LENGTH 33–37 cm **WINGSPAN** 71–79 cm **WEIGHT** no data

IDENTIFICATION SUMMARY Small compact gray buteonines. They are overall gray, darker above and paler below, and have red eyes and orange ceres, lores, and legs. Juveniles are similar to adults in plumage, but show white barring on belly, flanks, and leg feathers and white tips on uppertail coverts. They are rare and local in Panamanian rain forest and do not soar. On perched hawks, primaries extend about halfway down the short tail.

TAXONOMY AND GEOGRAPHIC VARIATION Monotypic, with no geographical variation. Formerly included in the genus *Leucopternis*, but DNA studies show they differ and belong in their own genus (Amaral et al. 2009).

SIMILAR SPECIES (1) **Plumbeous Kite** (plate 8) perched adults are also overall gray, but are more slender and have dark ceres, shorter legs, and wingtips that extend well beyond their tail tips. In flight, they show rufous primaries on pointed wings.

(2) **Hook-billed Kite** (plate 4) adult males are also overall gray, but have a larger beak, shorter legs, paddle-shaped wings with dark undersides, and two wide white bands on undertail.

(3) **Snail Kite** (plate 7) adult males are also overall gray with orange faces and legs, but they have dark underwings and white bases to the tail.

STATUS AND DISTRIBUTION Rare and local in rain forests of c and e Panama.

HABITAT Occurs in tropical rain forest.

BEHAVIOR These hawks hunt for prey from perches inside the forest. They prey on frogs, land crabs, fish, and snakes, and presumably small mammals and insects. They occasionally sit on exposed snags above the forest, especially early in the morning.

Their active flight is direct, with rapid wing beats. They do not soar.

The usual vocalization is a high, thin "wheee-urr."

Plumbeous Hawk. Adult. Adults have unmarked blue-gray underparts and blackish upperwings. Black tail has one white band. Note red eyes, red-orangish cere and gape, and orangish legs. Panama. March 2005. WSC

Plumbeous Hawk. Adult. Blackish upperwings and uppertail contrast with gray head and back. Note paddle-shaped wings. Colombia. April 2007. JÖRGEN FYHR

MOLT Not studied. Presumably annual molts are complete. Juveniles molt directly into adult plumage.

DESCRIPTION Sexes are alike in plumage, and juvenile plumage is similar to that of adults. Females are barely larger than males. Eyes are red to brown-red, and ceres, lores, and legs are bright orange. **Adult.** Upperparts are dark gray, and head and undersides are a paler medium gray. Some adults (younger?) show faint white barring on leg feathers. Upperwings are dark gray, and underwings are white except for dark tips of remiges. Short black tail has a narrow white band. **Juvenile.** Almost like adult, but they show whitish barring on belly, flanks, and leg feathers; a narrower dark subterminal band on underwings; white tips on undertail coverts; and a second, narrower white tail band. Their eyes are amber.

UNUSUAL PLUMAGES No unusual plumages have been reported.

HYBRIDS No hybrids have been reported.

ETYMOLOGY *Cryptoleucopteryx* comes from the Greek *kruptos*, "hidden," and *leukopterux*, "white-winged," for their underwings.

REFERENCES Amaral et al. 2009.

SAVANNAH HAWK PLATE 14
Buteogallus meridionalis

AGUILUCHO COLORADO

LENGTH 48–60 cm   **WINGSPAN** 125–138 cm   **WEIGHT** 650–1,070 g

IDENTIFICATION SUMMARY Large, slender, long-legged buteonines of open wet grasslands in Panama. Adults and Basic II (second-plumage) hawks are mostly rufous and black, but juveniles are brownish with little rufous. They often appear rather long-necked. Wingtips extend well beyond the tail tip on perched hawks.

TAXONOMY AND GEOGRAPHIC VARIATION Monotypic, with no geographic variation. Most authorities have included this species in the genus *Buteogallus*, but some favor the monotypic genus *Heterospizias*.

SIMILAR SPECIES (1) **Black-collared Hawk** (plate 14) is also a mostly rufous buteonine, but has a white head, a black bar across the upper breast, and wider wings, and lacks the white tail band. Adults have black remiges.

Savannah Hawk. Adult. Adults are overall rufous with barred underparts. All appear lanky with long legs. Argentina. June 2006. WSC

169

Savannah Hawk. Adult. Adults have rufous wings with wide black tips of remiges. Brazil. August 2012. WSC

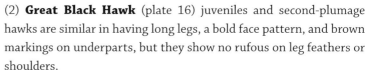

Savannah Hawk. Juvenile. Juveniles are overall buffy and brown, lacking rufous tones of adults. Venezuela. January 2006. WSC

(2) **Great Black Hawk** (plate 16) juveniles and second-plumage hawks are similar in having long legs, a bold face pattern, and brown markings on underparts, but they show no rufous on leg feathers or shoulders.

(3) **Harris's Hawk** (plate 23) is a somewhat similar buteonine, but has dark brown head and body, white tail coverts and white base of tail, and rufous restricted to wing coverts and leg feathers.

STATUS AND DISTRIBUTION Uncommon to local in open areas on the Pacific slope in Panama into sw Costa Rica and barely into e Panama. They have wandered north to Nicaragua.

HABITAT Occurs in open grasslands, especially savannahs and pastures.

BEHAVIOR They are general predators and capture a variety of prey, including small mammals, lizards, snakes, frogs, toads, crabs, eels, fish, and birds. They hunt from low perches, by walking on the ground, and from low flights. They regularly gather at fires and follow plowing tractors.

Active flight is with strong wing beats of somewhat cupped wings. They soar regularly with wings held level and glide with wings cupped. Display flights include mutual soaring, often with vocalizations and legs dangling. They build a stick nest in a variety of trees, including palms and mangroves, and they are silent except when breeding.

MOLT Not studied. Apparently the annual molt of body, coverts, and tail feathers is complete, but not all primaries and secondaries are replaced annually, the former result-ing in wave molt.

DESCRIPTION Sexes are alike in plumage, but females are somewhat larger. Adult plumage is quite distinctive, but juvenile plumage is somewhat different. Second plumage is intermediate. Legs are yellow to orange-yellow. **Adult.** Overall rufous except for dark back that shows a grayish cast, black tips of the remiges, and black tail with a white band. Rufous underparts are narrowly barred dark rufous. Note gray cheeks and amber eyes. **Basic II (second plumage).** Second-plumage hawks are similar to adults, and also appear overall rufous, but they show face pattern of rufous crown, buffy superciliaries, and dark eye-lines, and show unbarred buffy areas on underparts. **Juvenile.** Buffy head shows narrow dark streaking, except for superciliaries, cheeks, and throat. Buffy underparts show dark brown streaking, heavier on the breast, and dark barring on leg feathers. They show rufous when perched only on shoulder and leg feathers. Pale underwings show buffy, faintly marked coverts and pale rufous remiges with a dark brown border. Pale undertail shows wide dark brown terminal band. Some show little rufous.

FINE POINTS They spend lots of time walking on the ground, making their very long legs noticeable.

UNUSUAL PLUMAGES No unusual plumages have been described.

HYBRIDS No hybrids have been described.

ETYMOLOGY *Buteogallus* is from the Latin *buteo*, "hawk," and *gallus*, "chicken"—literally, "chicken hawk." The species name, *meridionalis*, is from the Latin and means "southern."

Savannah Hawk. Second plumage. Basic II hawks are intermediate between juvenile and adult. Panama. March 2005.
ROGER AHLMAN

COMMON BLACK HAWK PLATES 15–16
Buteogallus anthracinus

AGUILUCHO CANGREJERO

LENGTH 51–56 cm **WINGSPAN** 102–128 cm **WEIGHT** 630–1,300 g

BELOW LEFT: Common Black Hawk. Adult. Note that the orange on the cere bleeds onto the base of the beak. Wingtips fall short of the tail tip. Panama. March 2005. WSC

BELOW RIGHT: Common Black Hawk. Adult. Wings are broad and paddle-shaped. Note the small white areas on the base of the outer primaries and that the legs extend just to the tail band. Arizona USA. March 2015. WSC

IDENTIFICATION SUMMARY Large, wide-winged buteonines usually found near water, especially coastal mangroves. Adult and juvenile plumages differ. The beak is not completely black, and the legs are fairly long. On perched hawks, the wingtips of adults almost reach the tail tips, while the juvenile's wingtips fall somewhat short of the tip of the longer tail.

TAXONOMY AND GEOGRAPHIC VARIATION *B. a. anthracinus* occurs throughout and mixes on the Pacific slope of e Panama with *B. a. subtilis*, which differs mainly by showing rufous in the remiges of adults. Latter was considered a separate species by some, but Clark (2007) presented arguments for inclusion as a subspecies of *B. anthracinus*.

SIMILAR SPECIES (1) **Great Black Hawk** (plates 15–16) is similar in all plumages; see under that species for distinctions.

(2) **Solitary Eagle** (plates 15–16) is similar in all plumages; see under that species for distinctions.

(3) **Zone-tailed Hawk** (plate 22) adult perched may also show one white tail band on undertail, but band on uppertail is gray. It also has black, not yellow, face skin and shows barring on fore edge of folded wing. In flight, Zone-tail has more slender wings, which are two-toned below and always held in a strong dihedral.

STATUS AND DISTRIBUTION Fairly common to common, but local on both slopes of Mexico from s Sonora and s Tamaulipas, including the coastal Yucatán Peninsula, to Panama, and an uncommon and local summer resident inland in n Sonora into the US.

HABITAT Occurs in a variety of lowland wet habitats, including mangroves, open swamps, marshes, and riparian forest, but also in hilly rivers and mountain streams.

BEHAVIOR They are closely associated with aquatic habitats. They prey chiefly on crabs, fish, frogs, crayfish, snakes, reptiles, and large insects and occasionally on birds and mammals; their diet varies with locality, and they take the prey that is easiest (e.g., crabs in

Common Black Hawk. Juvenile. Juveniles are overall brown and buffy, with heavily streaked underparts and irregular white tail bands. Arizona USA. August 1994. wsc

mangroves). They are still hunters, sitting quietly on a perch, often a low one, overlooking a stream or small river, searching for prey. They often wade into shallow water and chase after prey on foot. Calls of this vocal species are a series of staccato whistled notes, very different from calls of other buteonines.

Active flight is with medium-slow, strong wing beats. Soars with wings held level and tail completely fanned, and glides with wings held level. The display flight of the male is a series of undulating climbs and dives, often performed while the bird is dangling its feet and calling. The pair often fly together and flutter their wings, which are held in a strong dihedral.

MOLT Apparently the annual molt of body, coverts, and tail feathers is complete, but not all primaries and secondaries are replaced annually, the former resulting in wave molt. Juveniles molt slowly into adult plumage, beginning at around nine months of age.

DESCRIPTION Sexes are nearly alike in plumage; females are slightly larger than males, with much size overlap. Juveniles have a different plumage, a longer tail, and narrower wings. Lores are whitish on all.

Adult. Head, body, and wing coverts are slate-gray, often with a grayish cast. Adult females usually show whitish on lores that extends under their eyes. Slate-gray upperwings show a variable amount of grayish mottling and a wide black terminal band on remiges. Some adults, especially in Panama, show a variable amount of rufous wash on remiges due to intermixing with *B. a. subtilis*. Underwings are slate-gray except for small white marks at the base of the outer two or three primaries. Black tail has one wide white band

Common Black Hawk. Juvenile. Juveniles are overall brown and buffy, with heavily streaked underparts and irregular white tail bands. Wings are broad and paddle-shaped. Note pale primary panels. Arizona USA. September 2013.
SAM ANGEVINE

and a thin white terminal band and appears short because of wide wings. Eyes are dark brown. Cere, face skin, and legs are bright orange-yellow. Cere color bleeds onto the base of the dark beak. **Adult (Mangrove Black Hawk,** ***B. a.*** **subtilis).** Differs from nominate adults only in strong rufous wash on remiges and larger and a more noticeable white slash on undersides of wings at base of outer primaries. **Juvenile.** Top of head is dark brown with buffy streaking. Strong face pattern shows buffy superciliary line, dark eye-line, buffy cheek, and dark malar stripes that extend onto the sides of the upper breast. Eyes are medium brown. Dark brown back and upperwing coverts show white and buffy feather edging. Dark brown upperwings show buffy primary patches in flight. Underparts are buffy, with irregular black spotting and streaking, often heavier and forming a dark patch on sides of upper breast. Underwings are buffy, with a black wrist comma. Buffy undertail coverts have black barring. White tail has irregular narrow black banding and a wide, black terminal band. Leg feathers are buffy with fine black barring. Cere is greenish-yellow. Legs are yellow. Cere color bleeds onto the base of the dark beak.

FINE POINTS White mark on underside at base of outer primaries is most visible when adults are flying away from the observer, but may be faint or absent.

UNUSUAL PLUMAGES Partial albinism has been reported from South America for this species, in one case an immature with some white and some normally colored feathers. Adult and juvenile specimens in amelanistic plumage have been collected in Panama.

HYBRIDS No hybrids have been reported.

ETYMOLOGY *Buteogallus* is from the Latin *buteo*, "hawk," and *gallus*, "chicken"—literally, "chicken hawk." The species name, *anthracinus*, means "coal black" in Latin.

REFERENCES Clark 2007.

GREAT BLACK HAWK PLATES 15–16
Buteogallus urubitinga
ÁGUILA NEGRA AMERICANA

LENGTH 55–65 cm   **WINGSPAN** 115–130 cm   **WEIGHT** 960–1,400 g

IDENTIFICATION SUMMARY Large, long-legged, and long-tailed buteonines usually found in wet grasslands and edges of rain forest. In flight, they show relatively narrow wings with leading and trailing edges nearly parallel and a long tail. Adult plumage is overall slate-gray, but the two immature plumages differ. Bill is entirely black, and cere is yellow, unlike those of Common Black Hawk. Two subspecies: nominate in e Panama and *B. u. ridgwayi* otherwise throughout. Wingtips barely extend beyond folded secondaries on perched hawks, when their long legs are noticeable.

TAXONOMY AND GEOGRAPHIC VARIATION *B. u. ridgwayi* occurs from s Mexico to w Panama and *B. u. urubitinga* in e Panama. Adults differ in tail pattern and leg markings, but immatures are hardly separable.

SIMILAR SPECIES (1) **Common Black Hawk** (plates 15–16) adults are also overall slate-gray and have a black tail with a white band, but they have wider wings that show white slashes at base of outer primaries and shorter legs that extend only to the tail band on the undertail of flying hawks. Adults show orangish on the base of the beak; this is horn-colored in juveniles. Juveniles differ in having a shorter tail that is white with irregular black banding and have wide dark malar stripes that extend onto sides of upper breast. All show extremely longer primary projection when perched, with primaries almost reaching tail tips.

(2) **Solitary Eagle** (plates 15–16) is similar in all plumages; see under that species for distinctions.

STATUS AND DISTRIBUTION Uncommon to fairly common in lowlands and foothills on both slopes from Mexico (s Sonora and s Tamaulipas), including the Yucatán Peninsula, south to Panama.

HABITAT Occurs in a variety of forests and wet open fields, usually not far from water.

BEHAVIOR Great Black Hawks are still hunters, but they move perches regularly and often forage on the ground. They take a variety of larger prey, including snakes (some venomous), lizards, small mammals, birds, fish, amphibians, large insects, fruit, and even carrion.

Active flight is with slow, deliberate wing beats. Soars and glides with wings held level. They soar often.

They build a rather large stick nest in a variety of trees.

They are quite vocal, both when perched and in flight, giving a long, harsh, shrill "keeeeee."

OPPOSITE PAGE: Great Black Hawk. Adult. Adults north of Panama show white barring on their leg feathers and one white tail band Belize. March 2011.
RICHARD PAVEK

MOLT Apparently the annual molt of body, coverts, and tail feathers is complete, but not all primaries and secondaries are replaced annually, the former resulting in wave molt. Beginning at around nine months of age, juveniles molt slowly into Basic II (second plumage), which is similar, but with darker head and a much different tail.

DESCRIPTION Sexes are alike in plumage, but females are larger. Juvenile and second plumages are different. Beak is entirely black, and lores are whitish on all. Cere is yellow. **Adult.** Head, body, and wing and tail coverts are slate-gray. Wings are rather narrow, with front and back edges more parallel. Remiges show pale gray banding and somewhat darker terminal band, and lack white slashes at base of outer three primaries on undersides. On flying hawks, long legs extend beyond tail band, almost to tip. White uppertail coverts and barring on leg feathers are diagnostic. Long black tail has wide white band and second narrower band near the tail base. Eyes are dark brown. **Adult (nominate).** Like *B. u. ridgwayi*, but the beak shows horn color at base, entire base of the tail is white, and the legs lack white barring. **Juvenile.** Buffy head is lightly marked, lacking wide dark malar stripes. Dark brown upperparts show some buffy mottling. Buffy underparts are darkly spotted and streaked. Buffy leg feathers have narrow dark barring. Buffy wings show dark markings on the underwing coverts, narrow dark banding on the flight feathers, and a pale panel in the primaries. Long brown tail shows many narrow

Great Black Hawk. Adult. Wings show nearly parallel leading and trailing edges. Note long legs extend beyond the tail band and lack of white areas at bases of outer primaries. Belize. March 2011.

RYAN PHILLIPS—BRRI

Great Black Hawk. Juvenile. Juveniles have brown tails with dark brown banding. Note that primaries extend barely beyond folded secondaries. Venezuela. December 2005. WSC

dark bands. Uppertail coverts are white. Eyes are brown, becoming darker with age. **Basic II (second plumage).** Similar to juvenile, but head is more darkly marked, and shorter tail is white with irregular wide black banding. **Juvenile and Basic II (nominate).** Similar to *B. u. ridgwayi*, but base of beak shows horn-colored area.

FINE POINTS Short primary projection and white uppertail coverts separate them from other similar dark raptors.

UNUSUAL PLUMAGES No unusual plumages have been described.

HYBRIDS No hybrids have been described.

ETYMOLOGY *Buteogallus* is from the Latin *buteo*, "hawk," and *gallus*, "chicken"—literally, "chicken hawk." The species name, *urubitinga*, comes from the Tupi Indian (Brazil) name *urubutsin*, "vulture."

Great Black Hawk. Adult South America. Adults in Panama show white base of the tail instead of a tail band and lack white barring on leg feathers (not shown). French Guinea. August 2006.
MICHEL GIRAUD-AUDINE

179

SOLITARY EAGLE PLATES 15–16
Buteogallus solitarius

ÁGUILA SOLITARIA

LENGTH 63–78 cm   **WINGSPAN** 157–180 cm   **WEIGHT** (few data) 2–3 kg

IDENTIFICATION SUMMARY Huge, wide-winged buteonines that are rare and local in forested hills and mountains. Adults are overall slate-gray, but appear paler gray on upperparts in flight, when the tail barely extends beyond the wide, triangularly shaped wings and the long legs extend almost to the tail tip. Immatures are dark brown with darkly streaked underparts and distinctive, uniformly dark leg feathers and unbanded tail. Wingtips of perched eagles exceed (adults) or just reach (juveniles) tail tip.

TAXONOMY AND GEOGRAPHIC VARIATION Monotypic, with no geographic variation. Formerly in the genus *Harpyhaliaetus*, but recent DNA studies place this taxon between the black hawks. The alleged subspecies *B. s. sheffleri* is hardly different.

SIMILAR SPECIES (1) **Common Black Hawk** (plates 15–16) is similar in all plumages, but they are always much smaller with narrower wings. Flying hawks show tail extending far beyond wings and legs, barely reaching mid-tail. Adults show a white slash on the outer three primaries, and juveniles show barred leg feathers and a banded tail; their beak has a pale base.

Solitary Eagle. Adult. Adults have all dark beaks, and their wingtips extend to or just beyond tail tip. Belize. September 2009.

YERAY SEMINARIO

180

(2) **Great Black Hawk** (plates 15–16) is similar in all plumages, but they are always much smaller, with much narrower wings that have leading and trailing edges somewhat parallel. Juveniles show barred leg feathers and a banded tail. All have distinctive whitish uppertail coverts.

STATUS AND DISTRIBUTION Rare and local in hills and mountains throughout from n Mexico to e Panama.

HABITAT Occurs in a variety of forests in hills and mountains, from tropical rain forest to pine-oak forest. Immatures must necessarily disperse great distances to find suitable nesting habitats, so they are possible in almost any habitat.

BEHAVIOR They prey primarily on snakes, but take other prey, especially mammals and birds. They hunt from perches, dropping to the ground to capture prey.

Solitary Eagle. Adult. Adults have broad wings that are somewhat triangular. Underwings lack white areas at the base of primaries. Long legs extend almost to tail tip. Ecuador. January 2007. BRIAN SULLIVAN

Solitary Eagle. Juvenile. Juveniles are brown and buffy, often with dark brown on upper breasts. Leg feathers are uniformly dark brown. Mexico. February 2006.

RICH CECH

Powered flight is with slow, strong wing beats. They soar often and soar and glide with wings held level.

The few nests described are large stick nests in tall pine trees.

Adults often vocalize in flight, giving a series of whistled "klee" calls.

MOLT Not studied. Presumably they do not molt all of their remiges annually, resulting in primary wave molt. Apparently all rectrices are replaced annually. Juveniles begin molt into second plumage when about nine months of age. Adult plumage is attained after two or three annual molts.

DESCRIPTION Sexes are alike in plumage, with females larger. Beak is black, and eyes are dark brown. **Adult.** Head, body, and coverts are dark slate-gray. Long nape feathers, when erected, form a bushy crest. Remiges are a paler gray, usually lacking dark banding, but with a wide dark terminal band. Black tail shows one wide white band. Cere is yellow, and long, stout legs are yellow-orange. **Juvenile.** Buffy head has dark brown crown and dark eye-lines. Dark brown upperparts and upperwing coverts show some narrow pale feather edging. Buffy underparts show narrow dark streaking and two large dark brown spots on sides of upper breast. Primaries are buffy below and brown above, with wide dark brown tips on outer ones. Secondaries are pale brown below with faint banding and brown above with dark tips. Buffy underwing coverts are lightly marked. Leg feathers are uniformly dark brown. Long tail is dark above and buffy below; tail has a dusky terminal band and can show faint narrow dark banding. Cere is dull yellow, and legs are yellow. **Basic II (second plumage).** Similar to juveniles, but new inner primaries and some secondaries have wide dark terminal bands. Head shows more black areas. New tail is shorter and pale gray below and medium gray above with a wide black terminal band. **Basic III.** Mostly adult-like, but overall dark brownish-black with buffy areas on the head and breast and underwing coverts. **Note.** Solitary Eagles' wing and tail shapes are not shown correctly in most bird field guides; see Clark et al. 2006.

FINE POINTS Solitary Eagles are the same size as King Vultures, whereas the black hawks are the size of Black Vultures.

UNUSUAL PLUMAGES No unusual plumages have been described.

HYBRIDS No hybrids have been described.

ETYMOLOGY *Buteogallus* is from the Latin *buteo*, "hawk," and *gallus*, "chicken"—literally, "chicken hawk." The species name, *solitarius*, comes from the Latin *solitarius*, "alone."

REFERENCES Clark et al. 2006.

BARRED HAWK PLATE 13
Morphnarchus princeps

AGUILUCHO PECHO NEGRO

LENGTH 49–57 cm   **WINGSPAN** 95–120 cm   **WEIGHT** 900–1,150 g

IDENTIFICATION SUMMARY Distinctive large buteonines of the mountains of Costa Rica and w Panama. All appear pale below with dark head and breast and dark, square-tipped tail with narrow white band. Adult males are smaller than females and a bit paler gray on the head and upperparts. Juveniles are similar but darker. Adults have blue-gray to horn-colored beak and orange ceres and lores. Juveniles have a dark beak and yellow cere and lores. Wingtips of both fall short of tail tip on perched hawks.

TAXONOMY AND GEOGRAPHIC VARIATION Monotypic, with no geographical variation. Formerly included in the genus *Leucopternis*, but recent DNA analyses show that this and other species belong in other genera (Amaral et al. 2009).

SIMILAR SPECIES No other raptor in its range is similar.

STATUS AND DISTRIBUTION Uncommon to fairly common in the highlands of Costa Rica and rare to uncommon in the mountains of w Panama.

HABITAT Occurs in montane forests and adjacent open lands.

BEHAVIOR Perch hunters, sitting quietly in trees inside or on edges of forest. When prey is sighted, they stoop to the ground to capture it. They take a variety of prey, including lizards, snakes, frogs, crabs, small mammals, and large insects. They are not often seen perched.

They soar regularly, often quite high, with wings held in a slight dihedral, but glide with wings level.

Vocalizations are heard mostly from soaring hawks and include a long "wheeyoor" and a series of "weeps."

MOLT Not studied. Annual molts of adult are apparently complete. Juveniles molt directly into adult plumage, beginning within a year of hatching.

DESCRIPTION Adult males are smaller than females and often a bit paler gray on the head, breast, and upperparts. Juveniles are similar but darker. Eyes are dark brown. **Adults.** Head, breast, and upperparts and barring on the underparts are dark gray, with males often a paler gray. Underwings are pale gray with fine dark barring on the coverts and remiges and black on the outer primaries. Short black tail shows one narrow white band. Adults have a blue-gray to horn-colored beak, orange cere and lores, and orange-yellow legs. **Juveniles** are similar to adults, but are slate-gray on head, breast, and upperparts and show slate-gray barring on underparts and have some narrow white feather edges on upperwing

Barred Hawk. Adult. Adults have dark gray head and breast, white belly, and whitish underwings narrowly barred gray. Ecuador. December 2008.
ROGER AHLMAN

Barred Hawk. Juvenile. Juveniles are like adults, but with darker slate-gray head and upperparts and yellow cere and gape. Costa Rica. March 2011.
CÉSAR SANCHEZ

coverts and back. Juveniles have a dark beak and yellow cere, lores, and legs.

UNUSUAL PLUMAGES No unusual plumages have been described.

HYBRIDS No hybrids have been described.

ETYMOLOGY *Morphnarchus* comes from the Greek *morphnos*, "dusky" or "dark," and *archus*, "leader" or "chief." The species name, *princeps*, is Latin for "chief" or "leader."

REFERENCES Amaral et al. 2009.

ROADSIDE HAWK PLATE 12

Rupornis magnirostris

GAVILÁN CAMINERO

LENGTH 33–38 cm   **WINGSPAN** 72–79 cm   **WEIGHT** 230–440 g

OPPOSITE PAGE TOP:
Roadside Hawk. Adult.
Rupornis m. petulans.
Adults are grayish.
Panama. March 2005.
WSC

IDENTIFICATION SUMMARY Small, chunky, buteonines with a barred belly. They are common and widespread throughout except for n Mexico. Adult plumages vary with locality, especially in the amount of rufous in the flight feathers and the color of the head, breast, and upperparts. Juvenile plumage is similar. Sexes are alike in plumage, with females somewhat larger. Wings are rather paddle-shaped. Buffy greater uppertail coverts form a U above tail base, visible on flying birds. Cere and legs are yellow-orange to orange. Wingtips reach just over halfway to tail tip on perched birds.

TAXONOMY AND GEOGRAPHIC VARIATION Three subspecies: *R. m. griseocauda* occurs from c Mexico to northwest Costa Rica, except for the Yucatán Peninsula and Belize, where *R. m. conspectus* occurs. *R. m. petulans* occurs from Costa Rica to Panama. Formerly included in the genus *Buteo*, but recent DNA studies have shown that this species is outside of that genus.

SIMILAR SPECIES (1) **Gray Hawk** (plate 17) juvenile is somewhat similar to juvenile, but has bold face pattern, dark eyes, tail bands of unequal width, and whiter, and more extensive U above tail base, and lacks streaked bib on underparts.

Roadside Hawk.
Adult. *Rupornis m.
griseocauda.* Adults are
brownish-gray. Belize.
December 2004. WSC

186

(2) **Broad-winged Hawk** (plate 18) is somewhat similar, but has more pointed wings, tail bands of unequal width, and dark eyes, and lacks bib on underparts.

(3) **Red-shouldered Hawk** (plate 17) is somewhat similar, but has crescent-shaped wing panels, tail bands of unequal width, and dark eyes, and lacks bib on underparts.

(4) **Sharp-shinned Hawk and Cooper's Hawk** (plates 10–11) are somewhat similar, but have shorter, more rounded wings and lack bib on underparts.

(5) **Hook-billed Kite** (plate 4) adults are somewhat similar and also have white eyes, but have a large hooked beak, collars on hind neck (not on adult male), rufous color in primaries (adult female), and underparts entirely barred.

(6) **Gray-lined Hawk** (plate 17) juveniles are somewhat similar, but they show dark brown blobs on underparts, not streaking or barring, and have white throats and underwing coverts.

STATUS AND DISTRIBUTION Common and widespread in Mexico on the Caribbean slope from s Tamaulipas south and on the Pacific slope from Jalisco south, the entire Yucatán Peninsula, and all of Central America except for mountains.

HABITAT Occurs at edges and open areas in a variety of lowland and foothill forests, especially humid ones, but not in pine-oak highland forests.

BEHAVIOR They perch conspicuously in open places, often on poles and wires along roadways. They are a general feeder, eating mainly insects and reptiles, but also rodents and sometimes birds. They hunt from perches and glide down to the ground to capture prey.

Active flight is accipiter-like, with three to five rapid, stiff, shallow wing beats followed by a glide. Soars with wings level; soaring is usually not very high, most often just above the forest. Glides with wings somewhat cupped, with wrists up and wingtips down. They do not hover. Pair in courtship flights glide together with fluttering wings held in a V, with tail closed and calling loudly.

Roadside Hawk. Adult. *Rupornis m. petulans.* Adults show lots of rufous in the remiges. Costa Rica. April 2006. WSC

They build stick nests in large trees.

Voice is a shrill scream that, once heard, is seldom forgotten.

MOLT The annual molts of adults are complete. Juveniles molt directly into adult plumage.

DESCRIPTION Adult plumages vary with locality, especially in the amount of rufous in the flight feathers and the color of the head, breast, and upperparts. Juvenile plumage is similar. Sexes are alike in plumage, with females somewhat larger. Cere and legs are yellow-orange to orange. **Adult (*griseocauda*).** Breast and head are dark gray-brown, and buffy belly is heavily banded dark rufous-brown. Buffy underwing coverts are lightly marked, and undersides of pale flight feathers show numerous narrow dark bands. Tail shows dark and pale bands of equal width (except on outer feathers), and buffy undertail coverts are unmarked. This race usually shows little or no rufous in the flight feathers. Eyes are *pale* yellow. **Adult (*petulans*).** Head and breast are gray, and upperparts are brownish-gray. Buffy belly is heavily barred rufous, and buffy leg feathers are narrowly barred rufous. Flight feathers show a noticeable rufous wash. Tail shows dark and pale bands of equal width. Some adults show some to lots of rufous in tail bands, completely so in many in Panama. **Adult (*conspectus*).** Similar to adult *petulans*, but overall much paler. **Juvenile**. Similar to adult, but head is browner, with noticeable creamy superciliary lines and pale brown eyes, and upperparts are dark brown. Breast (bib) is heavily marked with whitish streaks. Juveniles usually show no rufous in the flight feathers. Tail has more and narrower bands than those of adult. **Juvenile (pale extreme).** Similar to other juveniles, but with narrower and less obvious breast streaking and belly barring. Outer tail feathers show narrower dark bands.

FINE POINTS Adults usually have four dark tail bands; juveniles have five. Outer primaries are barred to tips.

UNUSUAL PLUMAGES An amelanistic juvenile reported from Veracruz, Mexico, has some paler beige feathers on the upperwing coverts and tail.

HYBRIDS No hybrids have been reported.

ETYMOLOGY *Rupornis* comes from the Latin *rupes*, "rock," and the Greek *ornis*, "bird." The species name, *magnirostris*, is from the Latin *magni*, "large," and *rostrum*, "beak."

OPPOSITE PAGE: Roadside Hawk. Juvenile. Juveniles show brownish head and upperparts, streaking in the dark breast, and noticeable white superciliaries. Mexico. December 2004. GREG LASLEY

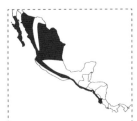

HARRIS'S HAWK PLATE 23

Parabuteo unicinctus

GAVILÁN MIXTO

LENGTH 46–59 cm **WINGSPREAD** 103–119 cm **WEIGHT** 568–1,203 g

IDENTIFICATION SUMMARY Long-legged, long-tailed, dark brown buteonines that inhabit deserts and mesquite woodlands. Large rufous upperwing patches are distinctive. Wings appear somewhat paddle-shaped in flight, and they glide with wings cupped. Juvenile plumage is similar to that of adult; both are dark and have rufous thighs and shoulder patches and white rumps. On perched birds, wingtips reach only halfway down the tail.

TAXONOMY AND GEOGRAPHIC VARIATION *P. u. harrisi* occurs throughout, with no variation.

SIMILAR SPECIES (1) **Dark-morph buteonines** (plates 18–22) lack rufous thighs and shoulder patches and white on both tail coverts and tail.

(2) **Snail Kite** (plate 7) adult male is similar in plumage and shape and glides on cupped wings, but lacks rufous patches and has distinctive thin, hooked beak.

STATUS AND DISTRIBUTION Fairly common in c and n Mexico and along a narrow dry belt down the Pacific coast to ne Costa Rica (Guanacaste). Some individuals wander far outside this area.

HABITAT Occurs in desert, thorn scrub, suburban areas, near agriculture, and in mesquite woodlands.

BEHAVIOR They prey mainly on mammals, especially rabbits, and birds, but also eat lizards and insects. They hunt on the wing and from perches. Cooperative hunting occurs more often in winter. Unlike many buteos, they don't hover. They perch more horizontally

Harris's Hawk. Adult. Overall dark brown, with rufous shoulders and leg feathers, and white tail coverts and base of tail. Texas USA. June 2015. WSC

than other raptors and are often seen in groups of up to a dozen in-dividuals, especially in winter.

Active flight is more energetic than that of buteos, with quick, shallow wing beats of cupped wings. Glides on cupped wings with wrists above body and wingtips pointed down. Soars with wings held more level.

They breed cooperatively, often with polygamy and nest helpers. A large stick nest is built in small to large trees and sometimes on power poles and cell towers.

Vocalization most often heard is alarm call, rendered as "iirrr."

MOLT Annual molt is complete, except for the remiges. Adults show primary wave molt. First-plumage adults show faint whitish mark-ings on primaries and have retained juvenile outer primaries and some secondaries.

DESCRIPTION Sexes are alike in plumage, but females are noticeably larger. Eyes are dark brown. Cere, lores, and legs are orange-yellow.

Adult. Head, body, and greater upperwing coverts are dark brown. Median and lesser upperwing coverts form rufous shoulder patches on perched birds. Underwings consist of rufous coverts and dark

Harris's Hawk. Adult. Underwings show rufous wing linings and dark remiges. Wide white tip of tail is noticeable. Texas USA. April 2014. WSC

gray flight feathers. Greater uppertail and undertail coverts are white. Black tail has wide white base and a noticeably wide white terminal band. **Juvenile.** Similar to adult, but dark back shows pale feather edges, and dark underparts are streaked white, heavier on belly. Underwings show rufous mottled coverts, finely barred pale gray secondaries, and whitish primaries with dark tips. Rufous thighs are barred with white; some show white thighs barred rufous. Dark brown tail shows white base from above, but appears light gray with fine dark bands below.

FINE POINTS First-plumage adults usually show faint whitish spotting on otherwise dark primaries.

UNUSUAL PLUMAGES Several specimens and live hawks were partial amelanistic. Several adults in south Texas had some white feathers.

HYBRIDS None have been reported from the wild. Captive-bred hybrids with several other raptor species appear bizarre.

ETYMOLOGY *Parabuteo* is from the Greek *para*, "beside or near," hence "similar to *Buteo*." The species name, *unicinctus*, is from the Latin *uni*, "once," and *cinctus*, "girdled," a reference to the white band at the base of the tail. This species is also called the Bay-winged Hawk, usually referring to the hawks in South America, although they may be a separate species. Named after Edward Harris, a friend of Audubon.

WHITE-TAILED HAWK PLATE 20

Geranoaetus albicaudatus

AGUILUCHO HOMBROS RUFOS

LENGTH 46–58 cm **WINGSPAN** 126–135 cm **WEIGHT** 700–1,350 g

IDENTIFICATION SUMMARY Large, long- and wide-winged buteonine with pointed wingtips and three immature plumages. Adults and juveniles appear quite different in shape and color. Wingtips greatly exceed the tail tip on perched older hawks, less so on juveniles.

TAXONOMY AND GEOGRAPHIC VARIATION Subspecies *hypospodius* occurs throughout. Two other subspecies occur in South America and have dark, rufous, and gray color morphs, not recorded for our subspecies. Formerly placed in the genus *Buteo*, but recent phylogenetic studies show that it is more closely related to two South American buteonines.

SIMILAR SPECIES (1) **Swainson's Hawk** (plate 19) are similar in having long wings with pointed wingtips, but show bright yellow

White-tailed Hawk. Adult. Adults have gray head and upperparts and white underparts, with rufous on upperwing coverts. Note wingtips extending far beyond tail tip. Texas USA. December 2013. WSC

ceres and underwings with coverts contrastingly paler than darker remiges, and primaries and secondaries the same color. Dark Swainson's are similar to juvenile White-taileds, but are smaller, have white undertail coverts, and lack pale cheek patches and white slash on the breast.

(2) **Turkey Vulture** (plate 1) in flight is similar to juvenile White-tailed, but lacks pale cheek patches, white U on uppertail coverts, and white slash on the breast.

(3) **Red-tailed Hawk** (plate 21) juveniles are similar to very pale juvenile White-tails that show a dark belly band, but Red-tails show dark patagial marks on underwings and rounded wingtips in flight, and when perched, their wingtips fall short of tail tip.

STATUS AND DISTRIBUTION Uncommon to fairly common, but local in Mexico in s Sonora, Tamaulipas, Durango, Zacatecas, Jalisco, Chiapas, and most of the Yucatán Peninsula; Belize; c Guatemala; Honduras; and local along the Pacific coasts of s Nicaragua, n and s Costa Rica, and w Panama.

HABITAT Occurs primarily in grassy open areas, sometimes also in agricultural fields, especially where sugarcane is being harvested or fields plowed.

BEHAVIOR They hunt both from perches and from the air by hovering and kiting. They gather in numbers at grass fires and when fields are being plowed or sugarcane is harvested; at these times numbers can be seen in flight and perched on the ground. Prey includes mammals, reptiles and amphibians, crustaceans, insects, and carrion.

White-tailed Hawk. Second plumage. Basic II hawks are similar to juveniles but have larger white breast patch and show rufous on upperwing coverts. Note wingtips extending far beyond tail tip. Texas USA. December 2013. WSC

They soar with wings in a strong dihedral and glide with wings in a modified dihedral. Powered flight is with slow, steady wing beats. They hover and kite often.

Large stick nests are usually placed in a lone small tree, yucca, or bush. They are not migratory, but non-breeding adults and immatures wander far afield.

MOLT Annual molt is not complete. Most body and covert feathers and rectrices are usually replaced, but not all of the remiges in the first and subsequent molts. First annual molt of juveniles begins when they are around nine months out of the nest. Adult plumage is attained after three or four molts.

DESCRIPTION Sexes are a bit different in plumages, and females are always larger. They have three immature plumages. Eyes are brown to dark brown, ceres are dull greenish, the dark beak has a bluish base, and the long legs are yellow. **Adult.** Head is gray, paler dove-gray on males, and throat is white, sometimes with gray streaking on females. Back and upperwing coverts are gray, with large rufous patches on wing coverts that become rufous shoulders on perched hawks. White underparts, lower back, and underwing coverts sometimes show faint narrow dark barring, heavier on females. Underwings show white coverts, dark primaries, and paler secondaries. White tail has a wide black subterminal band. **Juvenile.** Overall brownish-black with pale oval patches on the cheeks, white vertical slash or vertical oval on the breast, whitish tail coverts, and long grayish tail with indistinct narrow dusky banding; lacks a dark subterminal band. Undersides of remiges lack a dark subterminal band and are paler than dark coverts, resulting in a two-toned underwing, similar to that of Turkey Vulture. They usually don't show rufous on upperwing coverts. Some males are much paler, with lots of white on the underparts and underwing coverts, usually showing a dark belly band, similar to that of juvenile Red-tailed Hawk. **Basic II (second plumage).** Similar to juveniles, but are blacker and lack white cheek patches; white breast patch is larger and more rounded, and wings are broader through the secondaries. Remiges show a wide dark subterminal band, and shorter tail is grayer with a narrow dusky subterminal band. They usually have small rufous patches on the upperwing coverts. Males sometimes show rufous on the underwing coverts and belly. Some replace one to four rectrices a second time, usually the central ones, with feathers that are adult-like. **Basic III.** Similar to adult, but head, throat, and upperparts are slate-gray. Barring on flanks, legs, underwing coverts, and lower back is heavier and more distinct; these are often rufous on males.

FINE POINTS Bluish color of the base of the beak is stronger than the pale greenish of the cere, such that the cere seems bluish.

UNUSUAL PLUMAGES No unusual plumages have been described.

HYBRIDS No hybrids have been reported.

ETYMOLOGY *Geranoaetus* is from the Greek *geranos*, "crane," and *aetos*, "eagle." The species name, *albicaudatus*, is from the Latin *albus*, "white," and *caudatus*, "tail."

White-tailed Hawk. Third plumage. Basic III hawks are adult-like but have dark slate head and upperparts, with dark throats and heavy markings on their undersides. Texas USA. December 2013. wsc

WHITE HAWK PLATE 13
Pseudastur albicollis

AGUILUCHO BLANCO

LENGTH 48–58 cm **WINGSPAN** 98–117 cm **WEIGHT** 600–855 g

IDENTIFICATION SUMMARY Distinctive all-white, broad-winged large buteonine of lowland and foothill tropical rain forests. Beak is dark, cere is gray, lores are black, and legs are yellow. Two subspecies differ in the amount of black markings. Juvenile plumage is similar to adults'. Wingtips fall short of tail tip on perched hawks.

TAXONOMY AND GEOGRAPHIC VARIATION Two subspecies occur: *P. a. ghiesbreghti* is fairly common from s Mexico to w Nicaragua, and *P. a. costricensis* is fairly common from e Nicaragua to Panama and shows more black markings. Formerly included in the genus *Leucopternis*, but recent DNA analyses show that this and other species belong in other genera.

SIMILAR SPECIES (1) **Black and White Eagle** (plate 25) also appears whitish below, but has dark upperparts, narrower wings, and longer tail with multiple dark bands.

(2) **Short-tailed Hawk** (plate 18) also appears whitish below, but has dark head and upperparts, narrower wings that show dusky secondaries, and longer tail with multiple dark bands.

STATUS AND DISTRIBUTION Uncommon to fairly common on the Caribbean slope from s Mexico to Panama, but not in the Yucatán Peninsula; more local on the Pacific slope from s Mexico to Guatemala, and uncommon to fairly common from c Costa Rica to Panama. They prefer foothills over flat lowlands.

ABOVE: White Hawk. Adult *ghiesbreghti*. Perched adults are overall white, showing black only on wingtips. Belize. February 2006.
RYAN PHILLIPS—BRRI

White Hawk. Adults are overall white, showing black only on wingtips and tail band. Belize. April 2007.
RYAN PHILLIPS—BRRI

HABITAT Occurs in a variety of forests, especially tropical rain forest, but also subtropical and drier forests and adjacent treed pastures.

BEHAVIOR These perch hunters sit quietly in trees below the canopy or on forest edge. When prey is sighted, they stoop to the ground to capture it. They take a variety of prey, including lizards, snakes, frogs, crabs, small mammals, and large insects. They take some nestling and injured birds.

Active flight is with slow, soft, deliberate wing beats, rather moth-like. Soars and glides with wings level and pushed forward. Usually soars low over the forest, but sometimes quite high.

Vocalizations include a pig-like squeal: "sheeer" or "hee-EE-ooh."

MOLT Not studied. Annual molts of adult are apparently complete. Juveniles molt directly into adult plumage, beginning within a year of hatching.

DESCRIPTION Sexes are alike in plumage, but females are larger. Juvenile plumage is similar. Beak is

TOP: White Hawk. Adult *costaricensis*. Adults are overall white, showing more black on wings than *ghiesbreghti*. Costa Rica. April 2006. WSC

ABOVE: White Hawk. Juvenile. Juveniles are similar to adults, but have dark centers to secondaries. Belize. March 2010. RYAN PHILLIPS—BRRI

dark, cere is gray, lores are black, and legs are yellow. **Adult (ghiesbreghti).** Overall white with black tips to outermost primaries and greater primary upperwing coverts, and narrow black subterminal band on tail. Wide wings of adults make tail appear short. **Adult (costaricensis).** Similar to adult *gheisbreghti*, but secondaries are deep blue-gray with narrow black barring and show wide white tips; greater secondary wing coverts are deep blue-gray with narrow black barring and have wide white tips. Tips of greater coverts form a white line across each upperwing. Subterminal tail band is wider. **Juvenile (ghiesbreghti).** Like adult, but with narrower wings and some gray and black markings on wing coverts and secondaries. Juvenile tail appears longer than adult's. **Juvenile (costaricensis).** Similar to juvenile *ghiesbreghti*, but more heavily marked upperwing coverts and gray-barred secondaries. Tail appears longer than adults'. Compared to adult *costaricensis*, remiges are dark gray with narrow black barring and narrower white terminal band, and tail shows a narrower terminal band.

UNUSUAL PLUMAGES No unusual plumages have been described.

HYBRIDS No hybrids have been reported.

ETYMOLOGY *Pseudastur* comes from the Greek *pseud*, "false," and the Latin *astur*, "a kind of hawk." The species name, *albicollis*, comes from *albis*, "white," and *collis*, "necked."

SEMIPLUMBEOUS HAWK PLATE 12
Leucopternis semiplumbeus

GAVILÁN DORSO PLOMIZO

LENGTH 31–36 cm **WINGSPAN** 51–64 cm **WEIGHT** 215–368 g

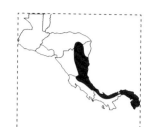

IDENTIFICATION SUMMARY Small compact gray and white bute-onines that are rare and local in rain forests of Panama and Costa Rica. They are dark gray above and white below, and have yellow eyes and orange ceres, lores, and legs. Juveniles are similar to adults in plumage, but show whitish streaking on the head, narrow dark breast streaking, and scaly markings on upperparts. Long black tail shows one (adult) or two (juvenile) narrow white bands. They do not soar. On perched hawks, primaries barely extend beyond folded secondaries.

TAXONOMY AND GEOGRAPHIC VARIATION Monotypic, with no geographical variation.

SIMILAR SPECIES (1) **Slaty-backed Forest-Falcon** (plate 28) is also gray above and white below, but has dark eyes, longer legs, and three white tail bands and lacks orange face and legs.

(2) **Short-tailed Hawk** (plate 18) light-morph is also gray above and white below, but their wingtips reach the tail tip, and they lack orange faces and legs.

STATUS AND DISTRIBUTION Rare to uncommon and local in rain forests of ec Honduras to c Nicaragua on the Caribbean slope to Panama and all of e Panama.

Semiplumbeous Hawk. Adult. Adults have gray heads and upperparts and white unmarked underparts. Note bright orange ceres, gapes, and legs. Costa Rica. March 2005.

NICK ATHANAS

199

HABITAT Occurs in tropical rain forest.

BEHAVIOR These still hunters perch in tropical forest and pounce on their prey of lizards, amphibians, small mammals, and birds.

Active flight is direct with rapid wing beats, with short periods of glide. They do not soar.

They build a nest in tall trees.

Usual vocalization is a long, high-pitched whistle.

MOLT Not studied. Presumably annual molts are complete. Juveniles molt directly into adult plumage.

DESCRIPTION Sexes are alike in plumage, and juvenile plumage is similar. Females are noticeably larger. Eyes are yellow, and ceres, lores, and legs are orange. **Adult.** Head and upperparts are gray, and underparts are unmarked white. Throat often appears dusky. Underwings show white coverts and remiges, the latter with some narrow dark banding. Long black tail has a narrow white band. **Juvenile.** Similar to adult, but with whitish streaking on the head, narrow dark breast streaking, upperparts that appear scaly, and a second white tail band.

UNUSUAL PLUMAGES No unusual plumages have been described.

HYBRIDS No hybrids have been described.

ETYMOLOGY *Leucopternis* comes from the Latin *leucos*, "white," and *pternis*, "a kind of hawk." The species name, *semiplumbeus*, is Latin from *semi*, "half," and *plumbeus*, "leaden" or "gray."

GRAY HAWK PLATE 17

Buteo plagiatus

GAVILÁN GRIS NORTEÑO

LENGTH 36–46 cm   **WINGSPREAD** 82–98 cm   **WEIGHT** 378–660 g

IDENTIFICATION SUMMARY Small, compact, accipiter-like buzzards, with relatively short wings with rounded tips. Pale underwings are relatively unmarked, except for dark border. Greater uppertail coverts form a distinctive white U above base of tail. They don't hover. They all show straight leading wing edges in flight. Two plumages: adult and juvenile. Adults are overall gray, and juvenile has a distinctive head showing white cheeks with bold dark eye-lines and malar stripes. Wingtips fall short of tail tip on perched birds.

TAXONOMY AND GEOGRAPHIC VARIATION Formerly considered conspecific with Gray-lined Hawk, *Buteo nitidus*; see Millsap et al. (2011). Amadon (1982) has assigned this taxon to the genus *Asturina*, but DNA analyses place it squarely in *Buteo*.

SIMILAR SPECIES (1) **Gray-lined Hawk** (plate 17) differs from Gray Hawk; see under that species for distinctions.

(2) **Broad-winged Hawk** (plate 18) juvenile is similar, but has darker head with brown, not white, cheeks; has streaked, not barred, leg

Gray Hawk. Adult. Dark gray upperparts lack pale barring. Mexico. October 2014. wsc

feathers; has four or fewer dark bands in tail; and lacks white U above base of tail. In flight, they show pointed wingtips. Adult Broad-wings have a rufous breast.

(3) **Roadside Hawk** (plate 12) perched appears somewhat similar; see under that species for distinctions.

(4) **Cooper's Hawk** (plate 11) juveniles are very similar in flight to juvenile Gray Hawks, with similar wing shapes, including straight leading edge of wings, and a long tail. But they show bold banding on remiges and rounded corners to the tail and lack white uppertail coverts.

(5) **Hook-billed Kite** (plate 4) adult male is similar to Gray Hawk adult, but has white eyes and paddle-shaped wings that appear dark below.

(6) **Red-shouldered Hawk** (plate 17) juveniles are somewhat similar; see under that species for distinctions.

STATUS AND DISTRIBUTION Uncommon to locally common resident from nw Costa Rica into s US, but absent from mountains and central plateau of Mexico, Sonora, and Baja California.

HABITAT Occurs in a variety of wooded habitats, usually near open areas, including riparian forest, open woodlands, and savannahs, and usually not far from water.

BEHAVIOR Bold, dashing raptors. Their relatively short wings and long tail allow them to maneuver in dense cover in pursuit of prey, mainly lizards and small birds, but occasionally insects and small mammals. They hunt mainly from perches, but occasionally from low glides, and by rapid stoops from a soar. They occasionally perch on poles or bare snags, especially in the non-breeding season.

Active flight is rapid and accipiter-like with quick wing beats followed by short glides on flat wings. Soars with wings level or in a slight dihedral, and glides on wings level or slightly depressed. They do not hover. Its courtship flights are similar to those of other buteos.

They build stick nests in tall trees.

Typical alarm call is a long drawn-out "wheeee."

MOLT Molt of body feathers is apparently complete annually. Juveniles begin molt into adult plumage when about nine months of age.

DESCRIPTION Sexes are alike in plumage. Females are larger than males, but with size overlap. Wings are relatively short for a buteo. Cere is orange-yellow to orange; legs are yellow. **Adult.** Head, back, and upperwing coverts are medium gray. Underparts are finely barred with white and medium gray. Whitish underwing is lightly barred gray. Undertail coverts are white. Black tail has two white bands, one wide and one narrow. Eyes are dark brown; cere is bright yellow. **Juvenile.** Top of head, back, and upperwing coverts are dark brown, with much rufous edging on coverts. Striking face pattern consists of a creamy superciliary line, a dark eye-line, white cheek and throat, and a dark malar stripe. Creamy to white underparts are heavily streaked with dark brown. Underwings appear whitish. Creamy leg feathers are barred dark brown. Long, medium brown tail has five or more dark brown bands, with the subterminal band the widest. Eyes are medium brown, and cere is bright yellow.

FINE POINTS When soaring, Gray Hawk's wingtips appear more rounded and tail longer than those of the Broad-winged Hawk. Gray's outer primaries are barred to the tips; Broad-wing's are not.

HYBRIDS An unsuccessful attempt at nesting with an adult Red-shouldered Hawk occurred in w Texas.

UNUSUAL PLUMAGES No unusual plumages have been reported.

ETYMOLOGY *Buteo* is Latin for "a kind of hawk or falcon." The species name, *plagiatus*, is from the Latin *plaga*, "stripe," referring to the adult's barred underparts.

REFERENCES Amadon 1982, Millsap et al. 2011.

BELOW LEFT: Gray Hawk. Adult. Overall pale gray below, with black tips on remiges. All soar and glide with leading edges of wings straight. Arizona USA. March 2009. NED HARRIS

BELOW RIGHT: Gray Hawk. Juvenile. Juveniles show distinctive face pattern and numerous dark tail bands, the outer two wider. Underwings are relatively unmarked. Arizona USA. August 2014. NED HARRIS

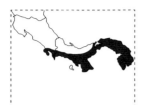

GRAY-LINED HAWK PLATE 17
Buteo nitidus

GAVILÁN GRIS SUREÑO

LENGTH 33–43 cm   **WINGSPREAD** 75–94 cm   **WEIGHT** 350–590 g

IDENTIFICATION SUMMARY Small, compact, accipiter-like buzzards, with short wings with rounded tips. They replace Gray Hawk in sw Costa Rica and Panama. Pale underwings are relatively unmarked and lack a dark border. They do not hover. Two plumages: adult and juvenile. Adults are overall gray and show barring on head and upperparts. Juveniles have heads that lack dark malar stripes and show buffy underparts with four distinct dark brown areas and dark and light tail bands of equal width. Wingtips extend to mid-tail on perched birds.

TAXONOMY AND GEOGRAPHIC VARIATION *B. n. costaricensis* occurs in Panama and sw Costa Rica, with no geographic variation. Species was formerly considered conspecific with Gray Hawk, *Buteo plagiatus*; see Millsap et al. (2011). Amadon (1982) has assigned this taxon to the genus *Asturina*, but DNA analyses place it squarely in *Buteo*.

SIMILAR SPECIES (1) **Gray Hawk** (plate 17) juvenile differs from juvenile Gray-lined in having white cheeks, narrow dark eye-stripes and malar streaks, whitish underparts narrowly streaked dark brown, and more and narrower dark tail bands; it lacks pale panel on upper primaries. Gray Hawk adults appear somewhat similar, but lack barring on crown and upperparts, show more tail bands and wider white uppertail coverts, and show a longer primary projection when perched.

(2) **Broad-winged Hawk** (plate 18) juvenile is similar, but has a darker head, shows dark malar stripes, and lacks white uppertail coverts and pale uppersides of primaries. In flight, they show pointed wingtips.

Gray-lined Hawk. Adult. Dark gray upperparts show pale barring. French Guinea. March 2013.

MICHEL GIRAUD-AUDINE

204

(3) **Roadside Hawk** (plate 12) perched appears similar, see under that species for distinctions.

STATUS AND DISTRIBUTION Uncommon and local resident in Panama and sw Costa Rica.

HABITAT Occurs in a variety of wooded habitats, usually near open areas, including riparian forest, open woodlands, and savannahs, and usually not far from water.

BEHAVIOR These are bold, dashing raptors. Their relatively short wings and long tail allow them to maneuver in dense cover in pursuit of prey, mainly lizards and small birds, but occasionally insects and small mammals. They hunt mainly from perches, but occasionally from low glides, and by rapid stoops from a soar. They occasionally perch on poles or bare snags, especially in the non-breeding season.

Active flight is with strong, stiff wing beats. Soars with wings level or with a slight dihedral, and glides on wings level or slightly depressed. They do not hover. Courtship flights are similar to those of other buteos.

They build stick nests in tall trees.

Alarm calls are shorter than those of Gray Hawks, with a drop in pitch in mid-call.

MOLT Molt of body feathers is apparently complete annually. Juveniles begin molt into adult plumage when about nine months of age.

DESCRIPTION Sexes are alike in plumage. Females are larger than males, but with size overlap. Wings are relatively short for a buteo. Primary projection is short. Cere is orange-yellow to orangish; legs are yellow. **Adult.** Pale head shows gray barring on the crown. Back

Gray-lined Hawk. Adult. Overall pale gray, lacking bold black tips on remiges. French Guinea. March 2013.
MICHEL GIRAUD-AUDINE

and upperwing coverts are gray with noticeable narrow pale gray barring. White underparts show narrow gray barring. Pale underwings lack a dark band on trailing wing edge. Dark uppertail coverts have white tips. Black tail has one wide white band and usually the hint of another narrower one. Eyes are medium to dark brown. **Juvenile.** Buffy head has dark streaks on the crown and dark eye-lines, but lacks dark malar stripes. Upperparts are brown, with some white or rufous edging. Creamy to buffy underparts are marked with dark brown blobs on the sides of the upper breast and usually on the flanks, and dark streaks that vary from heavily marked to lightly marked. Upperwings are two-toned: pale brown primaries contrast with darker secondaries. Underwings appear somewhat two-toned: dusky secondaries contrast with paler coverts and primaries. Creamy leg feathers are unmarked. Dark uppertail coverts have white tips. Tail has four pairs of light brown and dark brown bands of equal width. Eyes are pale gray-brown to pale brown.

UNUSUAL PLUMAGES No unusual plumages have been reported.

HYBRIDS No hybrids have been reported.

ETYMOLOGY *Buteo* is Latin for "a kind of hawk or falcon." The species name, *nitidus*, in Latin means "bright, shining."

REFERENCES Amadon 1982, Millsap et al. 2011.

Gray-lined Hawk. Juvenile. Juveniles have pale heads with prominent dark eye-line and dark blobs on pale undersides. Undertail shows equal-width dark and pale bands. French Guinea. September 2013.

MICHEL GIRAUD-AUDINE

RED-SHOULDERED HAWK PLATE 17
Buteo lineatus

GAVILÁN PECHIRROJO

LENGTH 38–47 cm **WINGSPREAD** 94–107 cm **WEIGHT** 460–930 g

IDENTIFICATION SUMMARY Medium-sized, long-legged, long-tailed, slender buteos usually found in wet woodlands or savannahs. Crescent-shaped wing panels, visible on upperwings and backlit underwings, are diagnostic. They do not hover. Two plumages: adult and juvenile. Adults with rufous underparts and shoulders and bold black and white remiges are distinctive. Juveniles are similar to other buzzards. Wingtips fall short of tail tip on perched birds.

TAXONOMY AND GEOGRAPHIC VARIATION Two subspecies occur: *B. l. texanus* is a winter visitor in e Mexico, and *B. l. elegans* is resident in most of Baja California.

SIMILAR SPECIES (1) **Broad-winged Hawk** (plate 18) juveniles perched are difficult to distinguish from perched juvenile Red-shoulders (see Fine Points). In flight, Broad-wings are smaller and have more pointed wings, with square, not crescent-shaped, pale wing panels. Adult tail has one wide white band (but may show another narrower white band near tail base). Adult underwing coverts are creamy, not rufous.

(2) **Gray Hawk** (plate 17) juvenile is similar in having brown upperparts and streaked underparts, but it has a noticeable head pattern of

Red-shouldered Hawk. Adult. Adults show rufous underparts and bold black and white on folded remiges. Florida USA. December 1995.
WSC

white cheeks set off by bold dark eye-lines and malar stripes, a bright yellow-orange cere, and narrow dark bands on a pale brown tail.

STATUS AND DISTRIBUTION Uncommon but local residents in all but s Baja California, Mexico, and rare to uncommon winter visitors to ne and c Mexico.

HABITAT Occurs in a variety of wooded and semi-open habitats, usually not far from water.

BEHAVIOR Perch hunters of wet woodlands, sitting quietly on inconspicuous perches searching for prey—mammals, birds, frogs and toads, snakes and lizards, and occasionally crustaceans, fish, and insects. In winter, they are seen in more open areas and regularly use more exposed perches, including telephone poles and wires.

Active flight is accipiter-like, with three to five quick, stiff, shallow wing beats, then a period of glide. They soar with wings level and with the leading edge pressed forward; they glide with wings bowed: wrists up, tips down. While gliding, birds flap occasionally. They do not hover. They are fond of soaring and often vocalize while in the air.

They are quite vocal, and birders should learn their distinctive call, which is especially evident during courtship.

MOLT Annual molts are usually complete, but a few adults will show a retained remex or two. Juveniles begin molt into adult plumage when about nine months of age.

Red-shouldered Hawk. Juvenile. Juveniles have brown upperparts and buffy underparts that are marked with dark diamonds or streaks. Note the three white bars on folded secondaries. Texas USA. December 2011. WSC

DESCRIPTION Sexes are alike in plumage, and females are a bit larger, with considerable size overlap. Two somewhat different plumage forms (especially juveniles) that correspond more or less with recognized subspecies: Texas (*texanus*) and California (*elegans*) are described. Adults and juveniles have different plumages. Legs and feet are pale yellow. **Texas adult.** Head is medium brown with tawny streaking. Back and greater upperwing coverts are dark brown with some rufous feather edges. Lesser and median upper-wing coverts are rufous and form the "red" shoulder of perched birds. Flight feathers are boldly barred above with black and white, not as boldly barred below, with a white, crescent-shaped panel near the black outer primary tips. Folded remiges and greater upperwing coverts are boldly black and white. Secondaries and inner primaries have white tips. Underwing appears two-toned: rufous coverts contrast with paler flight feathers. Underparts are rufous, varying from bright rufous to buffy-rufous, with narrow white barring; leg feathers are creamy with fine rufous barring. Black tail shows three or four narrow white bands. Eyes are dark brown. Cere is bright yellow. **Texas juvenile.** Head is medium brown, usually with buffy superciliary lines and dark brown malar stripes. Back is dark brown with some tawny mottling. Upperwing coverts are dark brown, with some tawny and whitish mottling and often a hint of the red shoulder. Folded secondaries show three rows of pale spots. Primaries are dark brown with a crescent-shaped tawny patch on the upper surface adjacent to black tips. Underparts are heavily marked with arrow-shaped markings, and leg feathers are usually barred, Underwing is uniform white to cream and shows a buffy crescent panel when backlit. Whitish undertail coverts are spotted with dark brown. Tail above is dark brown with four or six narrow pale bands; fewer bands are visible on the underside. Eyes are light to medium gray-brown. Cere is greenish-yellow. **California adult.** Similar to Texas adult, but always with brightly colored underparts. Head is more rufous than brown, almost the same color as the breast. Breast color is usually uniformly rufous, lacking whitish barring, and tail has fewer (two or three) and wider white bands. **California juvenile.** Different from Texas juvenile; more adult-like. Underparts are heavily marked, more barred than streaked. Underwing coverts are rufous, and so underwing appears two-toned like that of adult. Crescent-shaped patches on upperwings are whitish and similar to, but less bold than those of adult. Secondaries and inner primaries have white tips. Leg feathers and undertail coverts have thick rufous barring. Tail resembles that of adult, but is dark brown with whitish bands.

FINE POINTS Juvenile uppertail is dark with narrow pale bands; those of Broad-winged and Roadside Hawks are pale with dark bands. Juvenile secondaries have three pale spots on inner web that form three pale lines across folded secondaries of perched hawks. Secondaries of other juveniles are uniformly dark brown.

UNUSUAL PLUMAGES Amelanistic and partial albinism have been reported. One juvenile was all white with yellow eyes and faint tail banding.

HYBRIDS Hybrids with Red-tailed Hawks have been described. An unsuccessful attempt at nesting with an adult Gray Hawk occurred in Texas. A male Red-shouldered Hawk is producing juveniles with a female Common Black Hawk in California.

ETYMOLOGY "Red-shouldered" refers to the rufous patch on the upperwing coverts. *Buteo* is Latin for "a kind of hawk or falcon." The species name, *lineatus*, is Latin for "striped," a reference to the tail markings.

BROAD-WINGED HAWK PLATE 18
Buteo platypterus
AGUILUCHO ALAS ANCHAS

LENGTH 34–42 cm   **WINGSPREAD** 82–92 cm   **WEIGHT** 308–483 g

IDENTIFICATION SUMMARY Small, compact buzzards that have broad wings with pointed tips. Pale underwings are relatively unmarked, except for dark trailing edge. They do not hover. Four plumages: adult and juvenile of light and dark (less common) morphs. Wingtips fall short of tail tip on perched birds.

TAXONOMY AND GEOGRAPHIC VARIATION Only *B. p. platypterus* occurs in Mexico and Central America, with no geographic variation.

SIMILAR SPECIES (1) **Gray Hawk** (plate 17) juvenile is similar to juvenile Broad-wing, but has bolder black and white face pattern, barred leg feathers, and diagnostic white U above tail base.

(2) **Short-tailed Hawk** (plate 18) in flight is similar; see under that species for distinctions.

(3) **Roadside Hawk** (plate 12) perched appears similar; see under that species for distinctions.

(4) **Dark-morph buteos** (plates 21–22) of other species are larger, have different tail patterns, and do not have pointed wings.

Broad-winged Hawk. Adult. Adults have brown upperparts and buffy underparts marked with rufous-brown barring and a single pale tail band. Venezuela. December 2005. WSC

211

(5) **Swainson's Hawk** (plate 19) dark morph has dark gray, not silvery, flight feathers and contrastingly pale undertail coverts that contrast with dark belly, and wingtips reach tail tip on perched hawks.

STATUS AND DISTRIBUTION Fairly common to locally abundant migrants, and uncommon to fairly common winter visitors from c Mexico to Panama.

HABITAT They are seen on migration in large to small flocks. Preferred habitat is forest edge and trails and roads through forest, but sometimes in more open settings.

BEHAVIOR These still hunters perch on forest edges and sometimes on snags or poles in open fields. They prey on small mammals, birds, frogs and toads, snakes, and insects, which they capture after a short glide.

Active flight is with strong, stiff wing beats. Soars with wings level or with a slight dihedral, and glides on wings level or slightly depressed. They don't hover, but do soar regularly. On migration, small to large numbers of Broad-winged Hawks are often seen soaring together.

They are usually silent on migration and in winter, but they occasionally utter a two-noted, high-pitched whistle.

MOLT Molt of body feathers is apparently complete annually. However, not all remiges are replaced, resulting in primary wave molt. Wintering adults are in active molt, but molt is suspended during migration.

DESCRIPTION Sexes are alike in plumage. Females are larger than males, but with size overlap. Wings are relatively pointed for a buteo. Cere is greenish-yellow to yellow to orangish; legs are pale yellow.
Light-morph adult. Brown head shows dark malar stripes and a

white throat, usually with a dark mesial stripe. Back and upperwing coverts are uniform medium brown. Underparts are buffy, with variable amount of wide rufous-brown barring, heavier on the breast; sometimes breast is uniformly rufous-brown, forming a bib. Lightly marked underwings are white to cream, and primary and secondary tips are dark, forming dark band on trailing wing edge. Dark blackish-brown tail has one wide white band and another narrow one usually visible only when tail is fanned. Underwings are two-toned, with whitish flight feathers contrasting with dark brown coverts, more so in darker individuals. **Dark-morph adult.** Overall dark brown with variable dark rufous brown mottling in underparts to uniform dark brown to blackish. Eye color, remiges, and rectrices as in light-morph adult. **Light-morph juvenile.** Brown head shows pale brown superciliaries, cheeks, and throat and narrow dark malar stripes. Upperparts are brown, with some white or rufous edging. Creamy to buffy underparts are marked with brown streaks, varying from heavily marked to lightly marked, occasionally with clear breast and belly band to uncommonly unmarked or almost unmarked underparts. Underwings appear unmarked like those of adult, except that trailing edge band is not as wide or as dark. Creamy leg feathers are usually lightly streaked, rarely barred. Light brown tail has four or five dark brown bands; subterminal band is widest, or alternatively shows an accipiter-like pattern of dark and light bands of equal width. Eyes are pale gray-brown to pale brown. **Dark-morph juvenile.** Similar to dark-morph adult except that most birds have rufous and white mottling on underparts and underwing coverts. A few have uniformly dark brown bodies and underwings. Tail and eyes are identical to those of light-morph juvenile.

FINE POINTS Only three outer primaries are emarginated, unlike most other buteos, which have four.

UNUSUAL PLUMAGES There are reliable sightings of a completely albino adult and of a partial albino juvenile with many white feathers in its back and wings.

HYBRIDS No hybrids have been described.

ETYMOLOGY *Buteo* is Latin for "a kind of hawk or falcon." *Platypterus* is from the Greek *platys*, "broad," and *pteron*, "wing."

Broad-winged Hawk. Juvenile. Light. Juveniles' buffy underparts are marked with a variable amount of dark streaking and spotting. Texas USA, September 2007. WSC

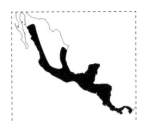

SHORT-TAILED HAWK PLATE 18
Buteo brachyurus
AGUILUCHO COLA CORTA

LENGTH 34–44 cm **WINGSPREAD** 86–103 cm **WEIGHT** 342–560 g

IDENTIFICATION SUMMARY Small, long-winged buteo, an aerial hunter of birds, usually seen in flight. Occurs in light and dark color morphs; light birds are more numerous, especially in s Central America. Best recognized by its unique flight behavior of facing into the wind and slowly moving forward while scanning below for prey; when they sight prey, they fold into a compact form and preform a spectacular stoop to the prey. Contrary to its name, the tail isn't particularly short. Wingtips reach tail tip on perched hawks.

TAXONOMY AND GEOGRAPHIC VARIATION *B. b. fuliginosus* occurs in Mexico and Central America, with no geographic variation. Nominate hawks in South America average smaller, with dark morph rare and with adults lacking the rufous breast patches.

Short-tailed Hawk. Adult light. Adults are not often seen perched. Note wingtips reach the tail tip. Arizona USA. February 2009. wsc

214

SIMILAR SPECIES (1) **Broad-winged Hawk** (plate 18) juvenile is similar to light-morph Short-tailed Hawk, but it almost always has streaked underparts, has uniformly pale underwings, and lacks dark cheeks. Wingtips do not reach tail tip on perched birds. Dark morph is similar to dark-morph Short-tailed Hawk, but is overall a paler brown and has remiges all the same silvery color and a different tail pattern. Juveniles are palely marked on the entire underparts, whereas juvenile dark Short-tailed Hawks are marked only on the belly.

(2) **Swainson's Hawk** (plate 19) light morph also has two-toned underwings, but its primaries and secondaries are the same darker gray color. They fly with wings in a strong dihedral and have a dark breast (adult) or usually more heavily streaked underparts (immature).

(4) **Swainson's Hawk** (plate 19) dark-morph hawks have contrastingly pale undertail coverts.

(5) **White-tailed Hawk** (plate 20) adult also shows dark cheeks and white throat, but is much larger and longer-winged and has a white tail with a dark black subterminal band. Their underwings are also two-toned, with darker primaries contrasting with paler secondaries, opposite of Short-tailed Hawk.

(6) **Zone-tailed Hawk** (plate 22) juvenile perched appears almost like perched dark Short-tailed Hawk, including having a white forehead, but is larger and has fewer white spots all over the body. Dark juvenile Short-tailed Hawk has more extensive white spotting, but this is restricted to the belly.

(7) **Black and White Eagle** (plate 25) is similar; see under that species for distinctions.

STATUS AND DISTRIBUTION Uncommon to fairly common residents, but local in some areas, throughout Mexico and Central America.

HABITAT Occurs in flight over a wide variety of habitats, both open and forested, from sea level to mountains, but not deserts.

BEHAVIOR These aerial hunters of small birds soar up to height, and face into the wind and hang stationary or move slowly forward on outstretched wings and fanned tail, with their heads down, searching the forest canopy or field below. When they spot prey, they fold their wings and stoop rapidly. Sometimes, to get a better look at potential prey, they lower themselves slowly on raised wings and then stoop. Often, after an unsuccessful stoop, they alternately flap and sail over the canopy, apparently hunting. They usually perch to rest and preen inconspicuously inside the canopy. They have been reported sitting on treetops with only their heads visible, apparently hunting, usually early in the morning.

Active flight is with stiff, strong wing beats. They soar with wings held level, but occasionally in a slight dihedral, and glide on level wings, often with tips turned up. In flight, they are best identified by their characteristic hunting behavior, kiting or moving forward slowly with head down and wings level with tips upturned.

They build a stick nest in tall trees, and they are usually silent, except when breeding, and then around the nest.

MOLT Apparently not well studied. Presumably annual molt is complete, with juveniles beginning molt when about 10 months of age.

DESCRIPTION Sexes are alike in plumage; females are larger. Juvenile plumages are similar to adults'. All show white foreheads. Wingtips reach tail tip on perched birds. Cere and legs are yellow. **Light-morph adult.** Head is dark brown. Solid dark cheeks and narrow white throat are noticeable in flight. Back is dark brown, and upperwing coverts are often a bit paler. Underparts are white, except for two small rufous patches on sides of upper breast. Underwings are two-toned: white coverts contrast with darker remiges. Primaries show an unbarred paler oval patch and are paler than secondaries. Tail above is brown with indistinct darker bands; tail below is whitish with a wide dark subterminal band and two or three indistinct, often incomplete, narrow dark bands. Eyes are dark brown. **Dark-morph adult.** Head, body, and wing coverts are dark brown. Eyes are dark brown. Underwings show an unbarred paler oval patch in the primaries, which are paler than secondaries. Tail is same as that of light-morph adult. **Light-morph juvenile.** Somewhat similar to light-morph adult, but has pale superciliaries, narrow pale streaks in the cheeks, and fine streaks on each side of the breast. Recently fledged birds show a rufous wash on breast. Undersides of remiges are paler than those of adults, resulting in less contrast with whitish coverts. Tail is brown above and pale below, with five or six narrow dark bands of equal width. Eyes are pale to medium brown. **Dark-morph juvenile.** Somewhat similar to dark-morph adult, but belly and underwing coverts are mottled with white; unmarked breast forms dark bib. Eyes and tail are like those of light-morph juveniles.

FINE POINTS Light-morph adults usually have small rufous areas on sides of upper breast, shown only by this subspecies.

UNUSUAL PLUMAGES No unusual plumages have been reported.

HYBRIDS An adult male tried to nest with an adult female Swainson's Hawk in Brownsville, Texas, including regular copulations, but the nesting was unsuccessful.

ETYMOLOGY *Buteo* is Latin for "a kind of hawk or falcon." *Brachyurus* is from the Greek *brachys*, "short," and *ours*, "tail."

Short-tailed Hawk. Juvenile dark. Dark juveniles are similar to dark adults but show whitish streaking on dark underparts. Wingtips reach tail tip on all. Florida USA. November 1996. SERGIO SEIPKE

sporadic winter records throughout

SWAINSON'S HAWK PLATE 19

Buteo swainsoni

AGUILUCHO LANGOSTERO

LENGTH 43–55 cm **WINGSPREAD** 120–137 cm **WEIGHT** 595–1,240 g

IDENTIFICATION SUMMARY Large slender buteos with long narrow wings, pointed wingtips, and a long tail. They are polymorphic, having light, rufous, and dark morphs. Two-toned underwing pattern is distinctive. White undertail coverts of dark- and rufous-morph hawks contrast with dark bellies. Wingtips reach or exceed tail tip on perched hawks. Numbers are seen on migration in spring and autumn at many locations. They winter locally in small numbers, especially in sw Mexico.

TAXONOMY AND GEOGRAPHIC VARIATION Monotypic: no geographic variation.

SIMILAR SPECIES (1) **Red-tailed Hawk** (plate 21) juvenile may appear similar to juvenile perched Swainson's as both have pale scapulars, but is distinguished by wingtips falling short of tail tip and underparts marked with a dark belly band.

(2) **Short-tailed Hawk** (plate 18) shares light-morph Swainson's Hawks' underwing pattern of light coverts and dark flight feathers, but its primaries are paler than secondaries, and it usually has unmarked underparts.

(3) **White-tailed Hawk** (plate 20) adults share light-morph Swainson's Hawks' underwing pattern of light coverts and dark flight feathers, but their secondaries are paler than primaries, and they have a white tail with a broad subterminal band. Juvenile White-

Swainson's Hawk. Adult dark-morph pair. Dark adults are overall dark but have white undertail coverts. Note wingtips extend a bit beyond the tail tip. Idaho USA. May 2000. wsc

tails have a similar wing shape, but wing coverts are darker than remiges.

(4) **Dark-morph buteos** of other species have dark undertail coverts and silvery flight feathers.

(5) **Prairie Falcons** (plate 31) are somewhat similar to pale juvenile Swainson's Hawks. But they have dark eyes and wingtips that fall short of the tail tip. In flight, they always show dark axillars.

STATUS AND DISTRIBUTION Uncommon to fairly common summer breeders in nc Mexico and a common to abundant migrant through our area. The migration path is rather narrow from Veracruz to South America. Small numbers winter along the Pacific slope of Mexico and in several areas of Central America, especially in s Honduras.

HABITAT Occurs in open grasslands and farmlands as a breeder, but passes over a variety of habitats on migration.

BEHAVIOR On migration, they are often seen in small to large flocks, flying or descending en masse to feed on grasshoppers. They are very kite-like in flight and hunt from perches as well as on the wing. They prey on insects, but also eat small mammals. They follow tractors and mowers, capturing disturbed rodents, insects, and even birds.

Active flight is light, with medium-speed, rather elastic wing beats. They soar with wings in a medium to strong dihedral and glide with wings in a modified dihedral. They hover and kite often, especially in strong winds.

ABOVE LEFT: Swainson's Hawk. Adult male. Adult males usually have rufous breasts. All show two-toned underwings: dark remiges contrast with pale wing linings. Utah USA. August 2013. JERRY LIGUORI

ABOVE RIGHT: Swainson's Hawk. Second plumage. Basic II hawks are somewhat like juveniles with dark blobs on either side of upper breast but have more heavily marked underparts and wider subterminal bands on remiges and rectrices. Idaho USA. May 2000. WSC

Swainson's Hawk. Second plumage. Basic II hawks are somewhat like juveniles but have more heavily marked underparts. Both show large dark areas on sides of upper breasts. Texas USA. September 2015. WSC

Large stick nests are built in small to large trees in open areas, and the usual vocalization is a drawn-out "keerrr," less wheezy than Red-tailed Hawk's.

MOLT Adults and Basic II (second-plumage) hawks molt during both summer and winter, but suspend molt while breeding and on migration. Juveniles begin their first molt soon after returning to the summer grounds and complete it on the winter grounds. They take two years to acquire adult plumage.

DESCRIPTION Sexes are almost alike in plumage, with females larger than males, but there is much size overlap. They have three color morphs: light, dark, and rufous. **Light-morph adult.** Brown head has a small white spot on the forehead and a large white throat patch. Back and upperwing coverts are brown, sometimes with a grayish cast, and contrast with darker brown remiges. Whitish greater uppertail coverts form a whitish U above tail base. Rufous to dark brown breast forms a bib that contrasts with lighter-colored belly, which is sometimes barred (more often on females). A few lightly colored adults have

incomplete bibs that are broken up by a pale area in the center of the breast. Underwings are two-toned: white coverts contrast with dark gray flight feathers. This is the only buteo with dark gray flight feathers. Light gray-brown tail has numerous narrow dark bands, with subterminal band wider. Eyes are dark brown. Cere and legs are yellow.
Dark-morph adult. Body and upperwing are dark grayish-brown. Underwing coverts vary from buffy to rufous to mottled dark brown to rarely entirely dark. Undertail coverts are pale, sometimes barred, and contrast with dark belly. **Rufous-morph adult.** Similar to dark-morph adult, but belly, underwing coverts, and leg feathers are medium to dark rufous. Breast is dark brown on females and rufous on males. Underwings are two-toned: rufous coverts contrast with dark gray flight feathers. White undertail coverts are barred with rufous and contrast with rufous belly. **Basic II (second plumage).** Similar to juvenile, but more heavily marked below, with new remiges darker and with wider dark subterminal bands. New rectrices also have wider dark subterminal bands. **Rufous- and dark-morph Basic II.** Similar to juvenile, but even darker below, with new remiges darker and with wider dark subterminal bands. New rectrices also have wider dark subterminal bands. **Light-morph juvenile.** Brown head is patterned: white forehead, streaked crown, buffy superciliary line, dark eye-line, buffy cheeks and throat, and dark malar stripes. Eyes are pale brown. Back and upperwing coverts are brown, with broad buffy feather edgings and a white U above the tail base. Underparts are white to creamy, lightly spotted to heavily streaked with dark brown, usually with two dark areas on the sides of the breast. Underwings are two-toned, but with less contrast than adult's. Some individuals' underwings appear uniformly pale and mottled. Brown tail shows numerous narrow dark brown bands. Cere is greenish-yellow to yellow. Legs are pale yellow.
Rufous- and dark-morph juveniles. Similar to light-morph juveniles, but underparts and underwing coverts are more heavily streaked and mottled, and back has little or no buff feather edging. Underwings are somewhat less two-toned because coverts are nearly as dark as flight feathers.

FINE POINTS Swainson's Hawks have only three notched primaries, a character shared with Broad-winged and White-tailed Hawks; all other buteos have four notched primaries.

UNUSUAL PLUMAGES Individuals with some white feathers have been reported.

HYBRIDS Hybrids with Rough-legged Hawks (Clark and Witt 2006) and Red-tailed Hawks (Hull et al. 2007), including Harlan's Hawk (Clark et al. 2005), have been reported.

ETYMOLOGY *Buteo* is Latin for "a kind of hawk or falcon." Common and scientific names are after William Swainson, English naturalist.

REFERENCES Clark et al. 2005, Clark and Witt 2006, Hull et al. 2007.

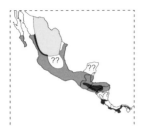

ZONE-TAILED HAWK PLATE 22
Buteo albonotatus
AGUILUCHO AURA

LENGTH 48–56 cm   **WINGSPAN** 121–140 cm   **WEIGHT** 610–1,080 g

IDENTIFICATION SUMMARY Black, slender-winged, long-tailed buteos. They are mimics of Turkey Vultures; the two species are alike in color, silhouette, and flight behavior. Sexes are almost alike in plumages, and females are noticeably larger. Juveniles are very similar to adults. On perched birds, wingtips reach tail tip. Cere and legs are bright yellow, and beak is entirely dark.

TAXONOMY AND GEOGRAPHIC VARIATION Monotypic, with no geographic variation.

SIMILAR SPECIES (1) **Turkey Vulture** (plate 1) is very similar, but has a smaller, unfeathered head and an unbanded tail, and lacks dark trailing edge on underwings. Both fly with frequent tilting.

(2) **Lesser Yellow-headed Vulture** (plate 1) is similar, but has a smaller, unfeathered head and an unbanded tail, and lacks dark trailing edge on underwings. Both fly with frequent tilting.

(3) **Common Black Hawk** (plates 15–16) adult has broader wings, all dark underwings with a small white mark at base of outer primaries, orange-yellow face skin and base of bill, and unbarred flight feathers. Its tail band is white above.

(4) **Dark-morph buteos** of other species are dark brown, not black.

(5) **White-tailed Hawk** (plate 20) juveniles are also black overall with similar underwings, but show more pointed wingtips and a white slash in the breast.

Zone-tailed Hawk. Adult. Adults have uniformly dark gray upperparts. Note that bands of folded uppertail are gray. Arizona USA. April 2015. SAM ANGEVINE

STATUS AND DISTRIBUTION Uncommon and local residents throughout in hilly and mountainous areas and in adjacent lowlands; more widespread in the non-breeding season. They are absent from much of the Yucatán Peninsula. Populations in mountains of n Mexico migrate to lower elevations in winter.

HABITAT Occurs in hilly and mountainous habitats throughout. They wander far afield in lowlands during the non-breeding season.

BEHAVIOR They hunt on the wing, mimicking Turkey Vultures so that they can approach prey close enough for capture before it can flee. They regularly join Turkey Vultures, usually flying below them, and showing their true identity and rapid flight only when potential prey is spotted. Then they stoop rapidly to snatch a bird, mammal, or lizard. They are often overlooked because of their similarity to Turkey Vultures; see Clark (2004b) for more on this mimicry.

Active flight is with medium-slow, flexible wing beats. They soar and glide like Turkey Vultures, with wings in a strong dihedral and with frequent rocking. They sometimes soar, buteo-like, with wings level and tail spread.

Talon-grappling between individuals of this species has been reported, usually in territorial defense. Vocalizations are typically buteo: a drawn-out, whistled "keeer." They can be very brave in defense of their nests and have on numerous occasions swooped at, and even soundly smacked, human intruders.

They build large stick nests in tall trees.

MOLT Annual molt of adults is complete, except for remiges, which show wave molt in the primaries. Juveniles molt into adult plumage beginning in the first spring.

DESCRIPTION Large, slender black buteo. Sexes are almost alike in plumage, but females are larger. Eyes are dark brown. Cere is bright yellow, and beak is completely black. Legs are orangish-yellow. **Adult.** Head, body, and wing coverts are black (but show a grayish bloom when seen in good light). Lores are pale gray. Flight feathers from above are somewhat two-toned, with primaries a bit paler than secondaries, and from below are light gray with heavy narrow dark gray barring and a broader black terminal band. Underwings appear two-toned: whitish remiges contrast with black coverts. Black tail has one wide band and either one (male) or two (female) narrow bands, which are gray above and white below. **Juvenile.** Similar to adult, but a somewhat browner black, with a variable amount of white spotting on body, heavier on underparts. Flight feathers are whiter below than those of adult, with narrow even-width dark barring and a narrower, less distinct terminal band of the adult. Tail is

Zone-tailed Hawk. Juvenile. Juveniles are similar to adults but have more finely banded remiges and pale undertails with narrow dark banding. They often show white spots on their underparts. Arizona USA. August 2014. SAM ANGEVINE

dark brown above, with black banding, and whitish below, with five to seven narrow dusky bands and a wide, dusky subterminal band.

FINE POINTS Underside of folded tail of adults shows one white band like that of Common Black Hawk, as their undertail coverts cover the narrower bands. When perched, Zone-tails show barring on leading edges of primaries, visible at a reasonable range and lacking on Black Hawks. The gray color on uppertail bands is on outer feather web only; bands are white on inner web, and uppertail bands appear white only when tail is widely fanned.

UNUSUAL PLUMAGES No unusual plumages have been reported.

HYBRIDS No hybrids have been reported.

ETYMOLOGY Named "zone-tailed" for the bands (zones) that are its tail markings. *Buteo* is Latin for "a kind of hawk or falcon." The species name, *albonotatus*, is from the Latin *albo* and *notatus*, "white" and "marked."

REFERENCES Clark 2004b

RED-TAILED HAWK PLATE 21
Buteo jamaicensis
AUGUILUCHO COLA ROJA

LENGTH 45–55 cm   **WINGSPAN** 120–141 cm   **WEIGHT** 750–1,500 g

Red-tailed Hawk. Adult *costaricensis*. Adults of this race show a reddish wash across the belly and underwing coverts. Costa Rica. April 2006. WSC

IDENTIFICATION SUMMARY Large polymorphic buteos resident in highlands and mountains, but a widespread but local winter visitor. They have three color morphs: light, rufous, and dark. The rufous tail of adults and the dark patagial marks on underwings are diagnostic field marks. Juveniles do not have a rufous tail. Adults in flight show a shorter tail and broader wings compared to juveniles, and thus have different flight silhouettes. On perched adults, wingtips reach or almost reach tail tip, but fall somewhat short on longer-tailed juveniles.

TAXONOMY AND GEOGRAPHIC VARIATION Three subspecies occur: *B. j. costaricensis* is widespread in the mountains from s Mexico to n Panama, including Belize. *B. j. fuertesi* is resident in parts of n Mexico, and *B. j. calurus* is resident on Baja California and a winter visitor throughout, south to Nicaragua. The alleged subspecies *B. j. kemseisi* (Guatemala to Nicaragua) and *B. j. hadropus* (s Mexico) are just paler variants of *costaricensis*, and typical hawks of the latter occur in the proposed ranges of the other two. Harlan's Hawk (taxon *harlani*) is treated separately, as there have been no published taxonomic justifications for treating it as a subspecies of Red-tailed Hawk.

SIMILAR SPECIES (1) **Light-morph buteos** of other species lack dark patagial marks on underwings.

(2) **Dark-morph buteos** of other species have non-rufous tails with different patterns.

(3) **Harlan's Hawk** (plate 32) juveniles are somewhat similar to juvenile rufous- and dark-morph hawks; see under that taxon for distinctions.

STATUS AND DISTRIBUTION Fairly common and widespread residents in the mountains from s Mexico to n Panama, including Belize, and a fairly common resident in parts of n Mexico. They are uncommon and local winter visitors throughout, south to Nicaragua, except for much of the Yucatán Peninsula.

HABITAT They occur as residents in mountainous and hilly forested and open areas, but are more widespread in a variety of open habitats during winter.

BEHAVIOR They are birds of both open and wooded areas, particularly wood edges, and they are often seen perched conspicuously on a treetop, a telephone pole, or other lookout while hunting. They prey mainly on rodents, but also on insects and their larvae, birds, reptiles, fish, and larger mammals, such as rabbits and squirrels.

Red-tailed Hawk. Adult *calurus*. Adults of this race have buffy underparts that are usually heavily marked. Dark patagial marks are shown by all light hawks. California USA. June 2012. RICHARD PAVEK

OPPOSITE PAGE: Red-tailed Hawk. Adult *fuertes*. Adults of this race have whitish underparts that are usually lightly marked. Oklahoma USA. November 2010. JIM LISH

They often pursue prey into dense brush, pirate prey from other raptors, and eat carrion. The most frequently heard vocalization is a long, wheezy "kkeeeeer," somewhat like the sound of escaping steam.

Active flight is with slow, steady, deep wing beats. They soar with wings raised slightly above the horizontal and glide with wings level or in a slight dihedral. They hover and kite on moderate wind, especially using deflection updrafts from wood edges, hills, and cliffs. Red-tails soar often.

Courtship displays are a series of steep dives and climbs and include glides by both adults together with their feet down. Talon-grappling has been reported and could be either courtship or aggressive display.

They build a large stick nest in a tall tree or cliff, but also on man-made structures, such as poles.

MOLT Annual molt of remiges is not complete, so adults show primary wave molt. First-plumage adults have retained one to three juvenile outer primaries and a few juvenile secondaries. Apparently, rectrix molt is complete. Juveniles begin molt into adult plumage beginning in their first spring.

DESCRIPTION Large polymorphic buteo. Plumages of the sexes are alike, but females are larger. *B. j. costaricensis* and *B. j. calurus* have three color morphs—light, rufous, and dark, but *B. j. fuertesi* occurs

only in light morph. Eyes of adults are medium to dark brown; those of juveniles are light gray-brown becoming pale yellow. Cere color is dull yellow to greenish-yellow. Legs vary from pale yellow to yellow.

Light-morph *costaricensis* adult. Head and upperparts are a uniform dark brown, except for whitish throat. Most show little or no rufous or buffy markings on the scapulars. Creamy underparts show a reddish wash on belly, undertail coverts, and leg feathers, which can fade to pale buff with time. Some adults, especially in s Mexico, show reddish barring on flanks and leg feathers. Creamy underwings show a reddish wash on coverts and dark patagial marks and narrow dark banding on remiges, with a wide dark terminal band. Uppertail coverts are rufous. Tail is rufous above with narrow dark subterminal band and can show hints of other narrow bands; tail is paler rufous below. **Light-morph *calurus* adult.** Head is brown with darker brown throat and cheeks, often with a golden crown and nape. Dark brown upperparts show buffy to rufous markings on the scapulars. Creamy underparts are variably marked with dark brown and rufous, especially on the belly. Leg feathers have brown to rufous barring. Rufous-buff underparts lack the reddish wash on the belly shown by adult *costaricensis*. Creamy underwings show dark patagial marks, some markings on buffy wing coverts, and narrow, wavy dark banding and wide dark terminal band on remiges. Uppertail coverts are rufous. Rufous tail has wide dark subterminal band and may have seven to ten additional, usually incomplete narrow dark bands. ***Fuertesi* adult.** Head is medium brown, with whitish throat and dark brown malar stripe. Back and upperwing coverts are dark brown, with white to pale buffy mottling often forming a V on back. Underparts, including undertail coverts and leg feathers, are creamy, often with a light rufous wash on sides of upper breast and usually with little or no belly band. Whitish underwing coverts show dark patagial marks and are sometimes lightly washed with rufous. Whitish tips of otherwise pale rufous or rufous-barred uppertail coverts form a white U above tail base. Rufous tail has a narrow dark brown subterminal band and can appear pinkish from below. **Dark-morph adult** (*costaricensis* and *calurus* are almost identical). Head, entire body, and upperwing and underwing coverts are dark chocolate brown to (rarely) brownish-black. Scapulars show no pale markings. Undersides of whitish primaries and secondaries are heavily barred, with a dark band on trailing edge of wings. Undertail coverts can be dark brown or dark rufous, somewhat paler than color of rest of underparts. Tail is rufous with a wide dark subterminal band and seven to eleven narrow dark bands. **Rufous-morph adult** (*costaricensis* and *calurus* are almost identical). Similar to dark-morph adult, but breast, underwing coverts, undertail coverts, and leg feathers are rich dark rufous. Dark patagial marks are darker than rufous coverts. Undertail coverts and leg feathers are heavily barred with dark brown. Wide belly band is dark chocolate brown and can show paler markings. **Light-morph *costaricensis* juvenile.** Head is medium brown, paler than adult's, with whitish superciliaries and on throat. Back and upperwing coverts are dark brown, with white mottling on scapulars forming a pale V on back. Upperwings are two-toned: primaries and coverts are somewhat paler brown and contrast with darker secondaries and coverts. White underparts show a belly band of dark markings. Recent fledglings have a rufous bloom on breast that fades quickly. White leg feathers are barred rufous. Whitish underwings show dark patagial marks, some dark markings on wing coverts, and narrow dark banding on remiges, with subterminal band paler and narrower than that of adult. Backlit underwings show square or trapezoid-shaped primary panels. Tail is light brown with many narrow dark brown bands of equal width, but occasionally the subterminal band is somewhat wider. **Light-morph *calurus* juvenile.** Similar to *costaricensis* juvenile, but throat is usually dark, upperparts are darker, belly band is more pronounced, and leg feathers have brown barring. ***Fuertesi* juvenile.** Similar to other juveniles, but overall paler and less heavily marked, always with white throat, faint markings on leg feathers, and narrow, dark belly band. Uppertail coverts can be whitish. **Dark-morph juvenile** (*costaricensis* and *calurus* are almost identical). Similar to dark-morph adult, but usually mottled with buff or rufous on underparts and underwing coverts, but can be a brown color overall without mottling. Tail is like that of light-morph *calurus* juvenile, but with wider dark bands. **Rufous-morph juvenile** (*costaricensis* and *calurus* are almost identical). They are not rufous, but are similar to light-morph *calurus* juvenile, except that it has heavily streaked breast, heavily mottled belly band and underwing coverts (but dark patagial mark still noticeable), and heavily barred leg feathers and undertail coverts. Tail is like that of light-morph *calurus* juvenile, but sometimes with wider bands.

UNUSUAL PLUMAGES Partial albinos, varying from almost all white birds to some with just a few white feathers, are fairly common and are reported from almost all areas. Most birds are from half white to mostly white. Mostly white individuals usually have a dark area on the nape. All individuals seen, reported, photographed, and in collections are adults. An amelanistic plumaged juvenile specimen is mostly cream-colored, with some faint rufous bars and streaks. An adult specimen, which appears normal in every other way, has a greenish gray, not rufous, tail.

HYBRIDS Hybrids of this species have been reported with Red-shouldered, Ferruginous, Swainson's, and Rough-legged Hawks.

ETYMOLOGY *Buteo* is Latin for "a kind of hawk or falcon." The species name, *jamaicensis*, is for Jamaica, where the first specimen was collected. Fuertes' Red-tailed Hawk was named for Louis Agassiz Fuertes, the bird artist. "Buzzard" is the proper name for these raptors. It comes from the same Latin root as *buteo* through Old French and Old English. European settlers of this continent mistakenly applied the name to vultures, confusing them for the darkish buteo that inhabits n Europe.

HARLAN'S HAWK PLATE 32
Buteo (jamaicensis) harlani

AUGUILUCHO COLA GRIS

LENGTH 45–55 cm   **WINGSPAN** 120–141 cm   **WEIGHT** 700–1,300 g

IDENTIFICATION SUMMARY A large buteo. Occurs in three color morphs: dark (85%), intermediate (5%), and light (10%). Adult and juvenile plumages are alike in body and coverts, but differ in remiges and rectrices. Most adults show pale to dark gray uppertails, with a variety of patterns, from mottling and speckling to banding, but color can include rufous, brown, and white. No two adults seem to have the same tail. Juvenile tails are also variable and usually have irregular and wide dark bands. Wingtips fall short of tail tip on perched Harlan's.

TAXONOMY AND GEOGRAPHIC VARIATION Monotypic, with no geographic variation. Harlan's has been treated as both a species and a subspecies of *B. jamaicensis* since its description by Audubon as a species in 1830. AOU declared it a subspecies in 1890 and 1972 without taxonomic justification and declared it a species in 1944 with justification. Best treated as a species.

SIMILAR SPECIES (1) **Ferruginous Hawk** (plate 22) dark-morph adults are also overall dark with white breast streaking; see under that species for distinctions.

(2) **Red-tailed Hawk** (plate 21) dark- and rufous-morph juveniles are similar to Harlan's juveniles, but they always show dark throats

Harlan's Hawk. Adult. Dark hawks usually but not always show whitish markings on their breasts. Undertail is usually whitish with a darker tip, but plumages are highly variable. Oklahoma USA. December 2009. JIM LISH

and completely dark outer primaries, and last dark band is next to the tail tip and lacks dark spikes.

STATUS AND DISTRIBUTION Rare winter visitors in several areas of n Mexico.

HABITAT Occurs in a variety of open areas, including grasslands, agriculture, pastures, wetter deserts, and edges of small copses of bare trees.

BEHAVIOR They are often seen perched conspicuously on a treetop, a telephone pole, or other lookout while hunting. They prey mainly on rodents, but also on larger mammals, such as rabbits and squirrels, and on snakes, frogs, and birds. They often pursue prey into dense brush, pirate prey from other raptors, and eat carrion. The most frequently heard vocalization is a long, wheezy "kkeeeeer," somewhat like the sound of escaping steam.

Active flight is with slow, steady, deep wing beats. They soar with wings raised slightly above the horizontal and glide with wings level or in a slight dihedral. They hover and kite on moderate wind, especially using deflection updrafts from wood edges and cliffs. They soar often. Compared to other raptors, they tend to be warier and flush at a greater distance when approached.

MOLT Annual molt of remiges is not complete, so adults show primary wave molt. First-plumage adults have retained one to three juvenile outer primaries and a few juvenile secondaries. Rectrix molt is usually complete. Juveniles molt into adult plumage beginning in their first spring.

DESCRIPTION Large polymorphic buteo. Sexes are alike in plumage, but females are larger. Occurs in three color morphs: dark (85%), intermediate (5%), and light (10%). Ceres are dull yellowish-green. **Dark-morph adult.** Head, entire body, and upperwing and underwing coverts are black, usually with much white speckling on breast and underwing coverts. A pair of white spots is often visible

Harlan's Hawk. Juvenile. Juveniles usually have barred tips on outer primaries. Tail bands are often wavy. Note the hour-glass markings on the tips of rectrices. Oklahoma USA. January 2009. JIM LISH

232

on the forewings of flying adults where the wings join the body (Liguori and Sullivan 2010). Underwings often show widely banded or unbanded but mottled silvery flight feathers, with a broad dark band on trailing edge. Tips of outer primaries are usually barred. Undertail coverts have some white mottling. Adult eyes are dark brown. **Adult intermediate morph.** Similar to dark-morph adult, but with white streaking uniformly on breast and belly. **Adult light-morph.** Dark head usually shows wide white superciliaries, white markings on the crown and cheeks, black malar stripes, and white throat. Uppersides are blackish with noticeable white crown streaks. Whitish underparts usually show some dark markings forming a belly band. Whitish underwings show white coverts with dark patagial marks and whitish remiges with barred outer primaries. Secondaries can be narrowly banded or mottled with no banding or a mix of these. Tail as on dark-morph adults. **Dark-morph juvenile.** Head, back, and upperwing coverts are blackish-brown, similar to those of dark-morph adult. Underparts are mostly blackish, with a variable amount of white streaking on breast (occasionally none) and some white mottling on belly. Underwing shows white mottling on blackish coverts and moderately barred flight feathers with a narrower and less dark band on trailing edge. Outer primaries usually have barred tips. Undertail coverts are barred black and white. Tail is similar to that of dark-morph juvenile Red-tail, but dark bands are usually wider and irregular. Separated in the field only with difficulty from juvenile rufous- and dark-morph Red-tails; some intergrades between these forms are encountered. Eyes are yellow. **Intermediate morph juvenile.** Similar to dark-morph juveniles, but with underparts uniformly streaked. **Light-morph juvenile.** Similar to adult light morph in body and coverts, but with juvenile remiges and rectrices. See Liguori and Sullivan (2010) for more detailed coverage of their field identification.

FINE POINTS Most juveniles show a dark spike or even a dark hourglass shape on the bottom of the last dark band on the tail and secondaries. These are not shown by any juvenile Red-tailed Hawk.

UNUSUAL PLUMAGES Partial albinos, varying from almost all white birds to some with just a few white feathers, have been reported. Most birds are from half white to mostly white. Mostly white individuals usually have a dark area on the nape. All partial-albino individuals seen, reported, photographed, and in collections are adults.

HYBRIDS No hybrids have been reported, but photographs exist of possible hybrids with Rough-legged Hawks; see Clark et al. (2005).

ETYMOLOGY *Buteo* is Latin for "a kind of hawk or falcon." The common name and species name are after Dr. Richard Harlan, who supported many of Audubon's bird-collecting expeditions.

REFERENCES Clark et al. 2005, Liguori and Sullivan 2010.

sporadic migration & winter records

FERRUGINOUS HAWK PLATE 22
Buteo regalis

AGUILUCHO FERRUGINOSO

LENGTH 50–66 cm **WINGSPAN** 134–152 cm **WEIGHT** 980–2,030 g

IDENTIFICATION SUMMARY Large buteos that have long, tapered wings, large heads and gapes, and robust chests. In flight, all show narrow black tips on undersides of outer primaries, and they lack dark band on trailing edges of wings. They have light and dark color morphs, and the sexes are alike in plumage, but females are separably larger than males. Juvenile plumage differs. Upperwings show whitish primary patches, but primary coverts are dark. Legs are tightly feathered down to toes. Wingtips fall short of tail tip on perched birds.

TAXONOMY AND GEOGRAPHIC VARIATION Monotypic.

SIMILAR SPECIES (1) **Red-tailed Hawk** (plate 21) light morph has dark patagial marks on underwing. Adult's rufous tail has a narrow dark subterminal band.

(2) **Harlan's Hawk** (plate 32) dark adults also have white breast streaking and grayish tail like dark adult Ferruginous Hawks, but have wider, conspicuous dark band on trailing edge of underwing, lack white wrist commas, and have boldly marked uppertail.

(3) **Rough-legged Hawk** (plate 22) dark-morph juvenile has same tail pattern as dark-morph juvenile Ferruginous, but lacks white wrist comma and has smaller head and gape.

Ferruginous Hawk. Adult. Adults have rufous on upperwing coverts. All show large gape, feathered tarsi, and wingtips that fall a bit short of tail tip. California USA. November 1996. wsc

Ferruginous Hawk. Juvenile. Juveniles have white unmarked underparts and brown upperwing coverts. All show large gape and feathered tarsi. Note lack of a dark malar. Idaho USA. May 1990. WSC

(4) **Dark-morph buteos** of other species lack white wrist comma and have different tail patterns.

STATUS AND DISTRIBUTION Rare winter visitors to open grassland areas of n Mexico and less often along the Veracruz coast.

HABITAT Open arid areas of n Mexico.

BEHAVIOR Adept flyers that hunt by coursing rapidly low over open ground, soaring at height; by hovering; or from a perch. Cooperative hunting of pairs has been reported. Their main prey is ground squirrels and jackrabbits, but they occasionally take other mammals and birds.

Active flight is with slow strong wing beats, much like that of a small eagle. Soars with wings held in medium to strong dihedral; glides with wings in a slight or modified dihedral. They hover occasionally.

MOLT Annual molt of adults is more or less complete, except that not all remiges are replaced annually, resulting in wave molt of the primaries. Juveniles molt into adult plumage beginning in their first spring, when they are about nine months old, and complete the molt by autumn.

DESCRIPTION Large, robust dimorphic buteo that has a huge gape. Occurs in light and dark color morphs. Plumages of adults are almost alike by sex, with females separably larger. All show whitish primary patches on upperwings, with dark primary coverts. Cere and feet are yellow. **Light-morph adult.** Whitish head shows creamy crown streaking and a dark eye-line. Cheeks are white with no dark malar stripe. Rufous back has narrow dark brown streaking. Upperwing

235

Ferruginous Hawk. Juvenile. Juveniles in flight show pale primary panels, pale base of uppertail, and dark line where secondaries reach the body. Arizona USA. December 2005.
WSC

coverts are more rufous than dark brown. Underparts are white, sometimes with belly band of rufous barring like Red-tailed Hawk. Underwings are white with a black wrist comma and, usually, rufous markings on coverts (more often on females). Leg feathers are rufous and barred with dark brown, but also sometimes white with rufous barring. Unbanded tail is white, light gray, light rufous, or some mixture of these, showing fine dark mottling on uppersides. Eyes are light to dark brown. **Light-morph juvenile.** White head has creamy crown streaking and a dark eye-line. Cheeks are white with no dark malar stripe. Back is dark brown, with little or no rufous. Upperwing coverts are dark brown, with some rufous feather edging. Underparts are white, with a narrow darkish line on each flank and sometimes a belly band of dark spots. Recently fledged birds have rufous wash on breast that fades by fall. Underwing is white with little or no dark markings and a black wrist comma. Leg feathers are white with black spots. Tail above is grayish-brown with basal third white and three or four incomplete dark bands; below it is silvery with dusky tip. Eyes are light brown. **Dark-morph adult.** Head, back, and upperwing coverts are dark brown, with rufous feather edges and rufous patches on the patagium and uppertail coverts. Underparts are dark brown, with a rufous patch and some white streaks on breast. Underwing is two-toned; silvery primaries contrast with dark coverts. Note white comma at wrist. Trailing edge of underwing has dusky border. Undertail coverts are dark rufous. Gray tail is unbanded; may have some dark mottling on uppersides, but appears whitish from below. Eyes as in light adult. **Dark-morph juvenile.** Head, body, and wing coverts are dark brown, sometimes with a hint of rufous on breast. Underwings are two-toned; silvery flight feathers contrast with dark coverts. Note white comma at wrist. Trailing edges of underwings show dusky border. Tail above is dark brown, with faint narrow dark bands; below it is silvery with dusky tip. Eyes are light brown.

FINE POINTS Their large gape is thought by researchers to be used for thermal regulation by allowing rapid air exchange, especially for nestlings, which may be in direct sunlight all day as many nests are on the ground with no shade.

UNUSUAL PLUMAGES No unusual plumages have been reported.

HYBRIDS Photos exist of a hybrid Ferruginous x Red-tailed Hawk.

ETYMOLOGY *Buteo* is Latin for "a kind of hawk or falcon." The species name, *regalis*, is Latin for "royal," a reference to the bird's large size. "Ferruginous" comes from the Latin *ferrugo*, "rust," for the rufous color in the adult plumage.

236

ROUGH-LEGGED HAWK PLATE 22

Buteo lagopus

AGUILUCHO ÁRTICO

LENGTH 46–59 cm   **WINGSPAN** 122–143 cm   **WEIGHT** 745–1,380 g

IDENTIFICATION SUMMARY Large, long-winged, and long-tailed buteos. They have light and dark color morphs, and adult plumages differ by sex. Juvenile plumage is similar to that of adult female. Second-plumage hawks are similar to adult females. Females are larger than males. Dark-morph individuals occur more frequently than in other *Buteo* species. Legs are loosely feathered completely to the toes. All but blackish adults show square black carpal patches on underwings. On perched hawks, wingtips reach tail tip. Cere is yellow, and toes are orange-yellow.

TAXONOMY AND GEOGRAPHIC VARIATION The North American race is *B. l. sancti-johannis.*, with no geographic variation. Other subspecies occur in Asia and Europe.

SIMILAR SPECIES (1) **Northern Harrier** (plate 9) also has white on uppertail coverts, but not on base of tail. They also lack dark carpal patches.

(2) **Ferruginous Hawk** (plate 22) dark-morph juveniles are very similar to Rough-leg juveniles; see under that species for distinctions.

(3) **Dark-morph buteos** of other species have different tail patterns.

STATUS AND DISTRIBUTION They are hawks of open country, breeding above the tree line on open tundra. The entire population moves south in winter and frequents open areas, such as farmlands, grasslands, marshes, and airports, occurring in small numbers in n Mexico.

HABITAT Occurs in open areas, such as grasslands, marshlands, and less arid deserts.

BEHAVIOR They hunt from low perches and hover in lighter winds. They prey almost exclusively on small to medium-sized mammals

Active flight is with slow, flexible wing beats. They soar with wings in a medium dihedral and glide with wings in a modified dihedral. They hover frequently; sometimes with deep wing beats, at other times with fluttering wings.

In winter, they form communal night roosts.

MOLT Annual molt of remiges is not complete, so adults show wave molt. Basic II (second-plumage) adults have retained one to three juvenile outer primaries and a few juvenile secondaries. Juveniles molt into Basic II plumage, beginning in their first spring. Apparently, rectrix molt is completed annually.

DESCRIPTION Large, dimorphic buteo. Occurs in light and dark color morphs. Plumages of adults differ by sex, with females separably larger. **Light-morph adult male.** Head is whitish to light brown, with streaked crown, dark eye-lines, dark malars, and a whitish nape. Back is dark gray-brown, with gray barring and tawny feather edging. Upperwings are dark and lack white primary patches. Whitish breast is heavily marked, appearing as a solid dark bib at a distance, and belly more lightly marked, sometimes clear white. Often a pale U-shaped area between breast and belly is noticeable. White flanks show dark barring, or dark flanks show white barring. White underwing coverts are heavily spotted dark brown. White leg feathers show dark barring. White tail shows wide black subterminal band and one to three narrow black bands, but some show a female-type tail. Adult eyes are dark brown. **Light-morph adult female.** Similar to adult male type, but flanks are uniformly dark, lacking any barring, and belly is more heavily marked than breast. They sometimes show a pale U between breast and belly. Upperparts are browner than adult males, lacking any gray barring. Underwings are like adult males', but with a buffy wash and more prominent carpal patch. White tail has wide dusky tip with narrower black subterminal band, but some show a tail like that of adult male. **Basic II (second plumage).** In their second plumage, both sexes appear much like adult females, but show incomplete remige molt and pale primary patches on upperwings. Eyes are medium brown, paler than those of adults. **Light-morph juvenile.** Head is creamy with brown streaking and a dark line behind eye. Upperparts are brown with some narrow pale feather edging. In flight, a whitish patch is visible on upperwing primaries, but primary coverts are dark. Creamy breast has some brown streaking, and belly has thick solid dark band. Underwings have clear creamy to white coverts with little mottling, prominent black carpal patch, and dusky trailing edge. Leg feathers are clear creamy, sometimes lightly spotted but never barred. White tail has wide dusky terminal band. Juvenile eyes are light brown. **Dark-morph adult male.** Some adult males are overall jet black except for silvery, heavily barred underside of flight feathers, which have dark tips forming dark trailing wing edge. There is often a whitish area on nape. Black tail has three or four narrow white bands. **Dark-morph adults.** Both sexes can be dark brown except for silvery, moderately barred underside of flight feathers.

Rough-legged Hawk. Adult female. Adult females have unmarked dark flanks. Adult males' flanks are barred. Washington USA. February 2002. WSC

238

Rough-legged Hawk. Juvenile. Juveniles show a wide uniformly dark belly band. All show black carpal patches. Oregon USA. February 2000. wsc

Underwings show rufous coverts, which contrast with black carpal patches. Tail is all dark above (dusky with darker tip) and silvery below with a dark terminal band, but some show narrow dark bands as well. **Dark-morph Basic II.** Almost like dark adults, but with incomplete molt of remiges and pale upper primary patches (Clark and Bloom 2005). **Dark-morph juvenile.** Overall color is medium to dark brown, often with rufous intermixed, sometimes with pale-colored head. In flight, a whitish patch is visible on upperwing primaries. Silvery flight feathers are lightly banded, with dusky tips forming a less boldly marked trailing edge of underwing than that of adult. Underwing coverts are dark brown or rufous; black carpal patch is noticeable. Tail above is dark, with faint darker bands; below it is silvery with dusky terminal band. **Note.** Some dark-morph juveniles and adult female types have pale-colored heads and whitish and tawny mottling on upper breast.

FINE POINTS They have small bills and feet and short gapes. Third-plumage adults can be distinguished by remnant of pale primary patches on upperwings (Clark and Bloom 2005).

UNUSUAL PLUMAGES No unusual plumages have been reported.

HYBRIDS Hybrids with Swainson's Hawks have been reported (Clark and Witt 2006, Clark et al. 2005). Some photos and specimens are thought to be hybrids with Harlan's Hawk.

ETYMOLOGY *Buteo* is Latin for "a kind of hawk or falcon." The species name, *lagopus*, is from the Greek *lagos*, "hare," and *pous*, "foot." Named "rough-legged" for the completely feathered legs.

REFERENCES Clark and Witt 2006, Clark et al. 2005, Clark and Bloom 2005.

CRESTED EAGLE PLATE 24

Morphnus guianensis

ÁGUILA CRESTADA

LENGTH 72–87 cm   **WINGSPAN** 130–152 cm   **WEIGHT** 1.4–2 kg

IDENTIFICATION SUMMARY Large, slender, long-legged and long-tailed forest eagles with bare tarsi. They show a single long crest of two feathers. Dark lores, and cere and beak form a shallow V on the front of the face, best seen head-on. They occur in light and dark morphs, and adult and juvenile plumages differ. They are rare and seldom seen due to their secretive behavior. They do not soar. Wings of flying eagles are short, broad, and rather paddle-shaped, and wingtips barely extend beyond folded secondaries on perched eagles.

TAXONOMY AND GEOGRAPHIC VARIATION Monotypic, with no geographic variation.

SIMILAR SPECIES (1) **Harpy Eagle** (plate 24) non-adults are somewhat similar. But they are even larger and bulkier and show two crests on the sides of the rear crown.

Crested Eagle. Adult. Adults have dark breast and paler belly that is narrowly barred. Long tail shows white bands. Black area in front of the eyes is distinctive. Belize. March 2013.
RICHARD KUENN-BRRI

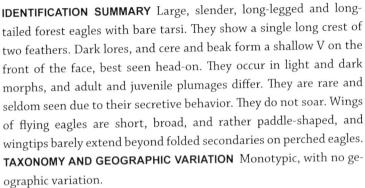

(2) **Other eagles** can also be overall whitish like juvenile Cresteds, but all have feathered tarsi and show different crests.

STATUS AND DISTRIBUTION Rare and local in lowland forest from s Mexico (absent on the Yucatán Peninsula) to n Nicaragua on the Caribbean slope and from Costa Rica to Panama on the Caribbean slope, and locally in sw Costa Rica and c and e Panama on the Pacific slope.

HABITAT Occurs in lowland tropical forest.

BEHAVIOR These perch hunters of rain forest prey on small to medium-sized mammals, as well as birds, reptiles, snakes, and tree frogs, which they attack by a stoop from a perch.

They do not soar, but have been reported to glide just above the forest. They glide with paddle-shaped wings held level. Powered flight is with strong wing beats.

They build a huge stick nest in canopy of large trees in forests.

MOLT Not studied. Most likely molt of remiges is not completed annually, resulting in primary wave molt.

DESCRIPTION Very large pale eagle with bare tarsi. Wings are paddle-shaped. All show black eye-rings, lores, and cere. Legs are yellow.

Light-morph adult. Pale to medium gray head shows darker crown feathers and two long dark crest feathers. Blackish back and upperwing coverts show narrow pale feather edges. Remiges are boldly banded black and white. Breast is medium to pale gray, and whitish belly shows pale to faint narrow rufous barring. Underwing and undertail coverts are white. Very long black tail shows two to four pale gray bands above that are white below. Eyes vary from gray to dark brown to amber. **Dark-morph adult.** Head, upperparts, and breast are blackish. White underparts and leg feathers show narrow black barring. White underwing coverts show some black barring; otherwise like light-morph adult, including the black face markings. **Light-morph juvenile.** Head and long crest are white, with dusky tips on the latter. Back and wing coverts are pale gray, and uppersides of blackish remiges are narrowly and irregularly banded white, with more white on undersides. White underparts are unmarked. Very long tail is pale gray with narrow, indistinct, broken black banding. Eyes are pale yellow. **Dark-morph juvenile.** Apparently much like light juvenile, becoming darker in subsequent plumages. **Light-morph older immature.** They appear somewhat intermediate between juvenile and adult, with dark tips to crest feathers, darker upperparts, hint of darker breast, and more distinct tail bands. Eyes gradually darken.

UNUSUAL PLUMAGES No unusual plumages have been described.

HYBRIDS No hybrids have been reported.

ETYMOLOGY *Morphnus* is Greek for a "kind of eagle or vulture." The species name, *guianensis*, is from Guyana, where the first specimen was collected.

ABOVE LEFT: Crested Eagle. Adult. Adults in flight show paddle-shaped wings that show strongly banded primaries. Brazil. June 2012. ANDREW WHITTACRE

ABOVE RIGHT: Crested Eagle. Adult dark. Dark adults are overall blackish, usually with white heavily barred belly. All show long narrow crest. Venezuela. October 1993. PETER BOESMAN

HARPY EAGLE PLATE 24
Harpia harpyja
ÁGUILA HARPÍA

LENGTH 89–108 cm **WINGSPAN** 175–200 cm **WEIGHT** 4–9 kg

Harpy Eagle. Adult. Adult's gray head has a split crest. Black breast band and unmarked white underparts are distinctive. Belize. February 2007.

IDENTIFICATION SUMMARY Huge, bulky, and powerful forest eagles that show two crests, one on each side of the head. Adults are black, white, and gray. Immature plumages show less black and gray. Wings of flying eagles are short, brown, and paddle-shaped, and the long tail is banded. Wingtips barely extend beyond folded secondaries on perched eagles. Bare legs are thick, and large feet are powerful. They do not soar.

Harpy Eagle. Adult. Adult's gray head, black band on breast, and heavily banded paddle-shaped wings are distinctive. Belize. March 2007.

TAXONOMY AND GEOGRAPHIC VARIATION Monotypic, with no geographic variation.

SIMILAR SPECIES (1) **Crested Eagle** (plate 24) is somewhat similar to non-adult Harpys; see under that species for distinctions.

STATUS AND DISTRIBUTION Rare and very local in remnant lowland forest on the Caribbean slope from s Mexico (but not the Yucatán Peninsula) to Panama and on the Caribbean slope of Panama and most of e Panama.

HABITAT Occurs in extensive lowland tropical forest.

BEHAVIOR These perch hunters in the upper and middle canopies of the rain forest prey on large mammals, especially sloths and monkeys, as well as large birds, lizards, and snakes, which they attack by a stoop from a perch.

They do not soar, but have been reported to glide just above the forest. They glide with paddle-shaped wings held level. Powered flight is with strong wing beats.

They build a huge stick nest in the canopy of large emergent trees in forests.

Vocalizations include a series of high, penetrating whistles and a loud, long, drawn-out whistle.

MOLT Not studied. Molt of remiges is not completed annually in such a large raptor; thus they show primary wave molt. Juveniles begin molt into second plumage while still dependent on their parents. Adult plumage is attained after three or four molts.

Harpy Eagle. Older immature. Immatures gradually acquire darker adult feathers. Split crest is distinctive. Belize. September 2006. RYAN PHILLIPS–BRRI

DESCRIPTION Huge forest eagle with bare tarsi. Sexes are alike in plumage, but females are much larger. Wings are broad and paddle-shaped. Dull yellow legs are thick and powerful. **Adult.** Gray head shows black crest feathers. Upperparts are black, and remiges are banded black and gray. White underparts show narrow black breast band. White leg feathers are narrowly barred black. Underwings show white and blackish coverts and boldly banded pale flight feathers. Long black tail shows two wide pale gray bands on uppersides and wide white bands on undersides. Lores and cere are blackish, and eyes are pale gray to brown. **Juvenile.** White head, crest, underwing coverts, and underparts are unmarked, except for a pale gray band across the breast. Pale gray back and upperwing coverts are somewhat mottled, and brownish remiges show indistinct narrow whitish banding and blacker tips of outer primaries. Long pale gray tail shows narrow distinct dark banding. Lores and cere are dark gray, and eyes are dark brown. **Younger immature.** Similar to juvenile, but head and breast are now pale gray, crest shows dusky tips, upperparts show some new dark feathers, and tail shows narrow dark bands. **Older immature.** Similar to adult, but with more whitish on dark upperparts and more and paler gray tail bands.

FINE POINTS Juveniles are dependent on adults for more than a year, so adults usually breed every other year.

UNUSUAL PLUMAGES No unusual plumages have been described.

HYBRIDS No hybrids have been described.

ETYMOLOGY *Harpia* and *harpyja* are from the Greek *harpe*, "bird of prey."

GOLDEN EAGLE PLATE 23

Aquila chrysaetos

ÁGUILA REAL

LENGTH 70–84 cm **WINGSPREAD** 185–220 cm **WEIGHT** 3–6.4 kg

IDENTIFICATION SUMMARY Large, dark, long-winged aerial eagles of hilly and mountainous areas of n Mexico. Golden nape and crown occurs in all plumages, varying from straw-yellow to deep orange-brown. Legs are completely feathered. Head projects less than half tail length on flying eagles. Bill and cere are tricolored, with beak tip dark, base horn-colored, and cere yellow.

TAXONOMY AND GEOGRAPHIC VARIATION The North American race is *A. c. canadensis*.

SIMILAR SPECIES (1) **Bald Eagle** (plate 3) immatures in flight are similar as large dark eagles; see under that species for distinctions.

(2) **Dark-morph buteonines** (plates 18–22) are much smaller and in flight show silvery flight feathers below, more strongly contrasting with darker underwing coverts.

(3) **Turkey Vulture** (plate 1) is smaller, flies with constant rocking or teetering, with wings in a strong dihedral, and has more contrasting two-toned underwing.

(4) **Common Black Hawk** juvenile (plate 15) shows white primary panels and whitish tail with dark terminal band, somewhat like

Golden Eagle. Adult. Adults lack white on remiges or rectrices, which do show some pale to medium-gray banding. All show golden hackles and feathered tarsi. Idaho USA. September 2009. WSC

juvenile Goldens, but has heavily streaked body, shorter wings and tail, and longer head extension.

STATUS AND DISTRIBUTION Rare and local in mountains and central plateau of n Mexico, some wandering to lower areas in autumn and winter. Their worldwide range includes the mountainous areas of Asia, Europe, the Middle East, n Africa, and North America.

HABITAT Occurs in forested and semiarid mountains, moving to lower elevations in winter.

BEHAVIOR They are agile flyers for their size and are able to take mammalian prey much larger than they are, including adult coyotes, deer, pronghorn antelope, fox, and bighorn sheep. They also capture large birds, such as Sage Grouse, Canada Geese, and even Whooping Cranes. Their favorite prey, however, is jackrabbits and other small mammals. In winter, they feed also on carrion. They hunt from perches and on the wing, when they glide slowly or kite on an updraft while searching the ground below. When prey is spotted, they stoop rapidly and directly. Mated adults often hunt as a pair. Most prey is captured on the ground, although they also take birds in flight.

Active flight is with slow wing beats. They soar with wings usually in a slight dihedral but sometimes level, and glide on level to slightly upraised wings, with wrists above body, primaries level with wrists, and wingtips upswept.

Golden Eagles usually nest on cliffs, but will use trees. They readily perch and even nest on man-made structures such as power poles.

MOLT Annual molt of body and flight feathers is not complete. Post-juvenile molt of the remiges is not complete, with only the inner three to six primaries and from none to half of the secondaries replaced, with new secondaries noticeably shorter. See Bloom and Clark (2001) for more on molts. Adult plumage is attained after four or five molts.

DESCRIPTION Sexes are similar in plumage, with females larger. Legs are feathered to the toes. Cere and gape are yellow, and beak is two-toned, horn-colored at base and dark on tip. Feet are yellow. **Adult.** Head is dark brown except for golden crown, upper cheeks, crest, and nape. Body and wing coverts are dark brown, with some dark rufous on underwing coverts and undertail coverts. Body plumage often appears mottled because new dark feathers contrast with old paler feathers. Tawny mottling on median secondary upperwing coverts form diagonal bars on inner upperwing. Flight feathers are gray-brown below and appear somewhat lighter than underwing coverts when seen in good light, when a dark subterminal band is noticeable. Faint wavy gray bands are visible in tail and flight feathers in good light. **Juvenile.** In the first plumage, head is like adults', and body and coverts are uniformly dark brown. Recently fledged birds may appear almost black. Upperwing coverts are uniformly dark brown, without tawny bars. White patches at base of inner primaries and outer secondaries are usually visible on flying birds from below: these are sometimes visible from above as smaller patches. Some juveniles have small white patches; some lack them completely. Tail has white base, including white edges, wide dark brown subterminal band, and narrow white terminal band. White on tail is more easily seen from above.

Older immatures. These plumages occur in the next three years and are similar to that of the juvenile; white patches in the flight feathers remain about the same size, but become progressively irregular. Tawny median upperwing coverts form diagonal bars on upperwings, visible on both flying and perched birds. Tail has progressively less white on base (usually with white edges) and some wavy gray lines in the dark tip. **Note.** Accurate determination of the age of non-adult Golden Eagles using only the amount of white in the wings and tail is not possible because of individual variation and the lack of variation in immatures with age.

FINE POINTS Golden Eagle nape does not change color throughout the eagles' life, but color varies among individuals. The tail of the adult male has fine wavy gray bands; that of the adult female has one wide and one narrow irregular wavy gray band.

UNUSUAL PLUMAGES A few cases of partial albinism have been reported.

HYBRIDS A hybrid with Imperial Eagle (*Aquila heliaca*) was reported from Europe.

ETYMOLOGY "Golden" refers to the color of the crown and nape feathers. *Aquila* is Latin for "eagle," and *chrysaetos* is from the Greek *chrysos*, "golden," and *aetos*, "eagle."

REFERENCES Bloom and Clark 2001.

Golden Eagle. Juvenile. Juveniles are overall uniformly dark. All show yellow cere, two-toned beak, and buffy feathered tarsi. Idaho USA. May 1995. WSC

BLACK HAWK-EAGLE PLATE 25
Spizaetus tyrannus
ÁGUILA CRESTUDA NEGRA

LENGTH 55–70 cm **WINGSPAN** 110–135 cm **WEIGHT** 900–1,300 g

IDENTIFICATION SUMMARY Medium-sized black forest eagle with cowl-like crest and feathered tarsi. Typical hawk-eagle with short, broad, paddle-shaped wings, feathered tarsi, and long tail. Adults are mostly black, second-plumage eagles are adult-like, but with more white, and juveniles are paler. Wingtips barely extend beyond folded secondaries on perched eagles.

TAXONOMY AND GEOGRAPHIC VARIATION *S. t. serus* occurs throughout Mexico and Central America, with no geographic variation.

SIMILAR SPECIES (1) **Hook-billed Kite** (plate 4) dark morph is similar. They are smaller, larger-beaked, and shorter-tailed, and have more paddle-shaped and darker wings that are less heavily marked and shorter bare tarsi.

(2) **Crested Eagle** (plate 24) dark morph is somewhat similar. But there is only one record of this morph in our area.

STATUS AND DISTRIBUTION Black Hawk-Eagles are rare to uncommon and local residents on the Caribbean slope in lowland and foothill areas from s Mexico including the Yucatán Peninsula, to Panama, and rare on the Pacific slope of s Mexico, and uncommon to fairly common residents on both slopes from s Nicaragua to Panama.

HABITAT Occurs in heavily and lightly forested areas, including second growth, in lowlands, foothills, and lower mountains.

BEHAVIOR Black Hawk-Eagles are still hunters, pursuing prey inside forests from perches. They make short stoops from perches, but also

Black Hawk-Eagle. Adult. Adults are overall black but with white spots on underparts and white barring on leg feathers. Note short cowl-like crest. Colombia. June 2013.

NICK ATHANAS

tail-chase escaping prey. They prey primarily on arboreal mammals and birds, but they also take large lizards and snakes.

They soar and glide with wings held level. Powered flight is with strong, quick wing beats. Display flights include high-altitude mutual soaring, undulating flight, and wing fluttering, often accompanied by vocalizations. They soar frequently, most often in late morning.

Large stick nests are built high in tall trees in forests.

MOLT Not studied. Presumably they do not molt all of their remiges annually, resulting in primary wave molt. Apparently all remiges are replaced annually. Juveniles begin molt into Basic II (second) plumage when about nine months of age.

DESCRIPTION Black typical hawk-eagle, with short, cowl-like crest and completely feathered tarsi. Sexes are alike in plumage, with females larger. Juvenile plumage differs, but Basic II (second) plumage is adult-like. Lores are gray, and cere is dull pale yellow to gray. **Adult.** Overall black, with white in raised cowl, white barring on uppertail coverts, leg feathers, and undertail coverts, narrowly and boldly barred black remiges, and long tail with black and mottled gray bands of equal width. Eyes are orangish. **Juvenile.** Brownish head shows bold white superciliaries, darker auriculars, and white throat. White extends onto center of breast, which is otherwise narrowly streaked dark brown. Whitish belly is boldly barred dark brown and black. Whitish leg feathers and undertail coverts are narrowly barred dark brown. Dark brown upperparts show narrow pale feather edges. Long tail shows dark brown and pale grayish tail bands of equal width. Eyes are yellow. **Basic II (second plumage).** Second-plumage eagles are adult-like, but with retained whitish areas on head and belly.

UNUSUAL PLUMAGES No unusual plumages have been described.

HYBRIDS No hybrids have been reported.

ETYMOLOGY *Spizaetus* is from the Greek *spizias*, "hawk," and *aetos*, "eagle." The species name, *tyrannus*, is from the Latin *tyrannus*, "tyrant," for its belligerent and dominant behavior.

BELOW LEFT: Black Hawk-Eagle. Adult. Adults show heavily banded paddle-shaped wings, long tail with black and white bands, and feathered tarsi. Belize. March 2007. RYAN PHILLIPS—BRRI

BELOW RIGHT: Black Hawk-Eagle. Juvenile. Juveniles are similar to adults but have pale heads, streaked breast, and barred belly. Belize. March 2011. BARRY ZIMMER

ORNATE HAWK-EAGLE PLATE 25
Spizaetus ornatus
ÁGUILA CRESTUDA REAL

LENGTH 56–67 cm **WINGSPAN** 107–130 cm **WEIGHT** 840–1,750 g

IDENTIFICATION SUMMARY Medium-sized, long-crested forest eagles with feathered tarsi. Adults have striking plumage, but juvenile plumage is overall whitish. Basic II (second) plumage is intermediate but adult-like. All in flight show rather paddle-shaped wings and a long tail. Wingtips barely extend beyond folded secondaries on perched eagles.

TAXONOMY AND GEOGRAPHIC VARIATION *S. o. vicarius* occurs in Mexico and Central America, with no geographical variation.

SIMILAR SPECIES (1) **Black and White Eagle** (plate 25) is similar to juvenile Ornate Hawk-Eagle; see under that species for distinctions.

(2) **Gray-headed Kite** (plate 4) paler juveniles are similar; see under that species for distinctions.

(3) **Gray-bellied Hawk** (plate 32) juveniles are similar to adult Ornate Hawk-Eagles; see under that species for distinctions.

(4) **Hook-billed Kite** (plate 4) juveniles in flight are similar to juvenile Ornate Hawk-Eagles, but they are smaller, have dark crowns and more heavily banded undersides of remiges, and lack barring on axillars.

Ornate Hawk-Eagle. Adult. Adults are distinctive, with long dark crest and black, white, and orange markings on head, neck, and breast. Belize. April 2006. ANTONIO HIDALGO

STATUS AND DISTRIBUTION Rare to uncommon and local residents on the Caribbean slope in lowland and foothill areas in s Mexico, including most of the Yucatán Peninsula, and on the Pacific slope, and rare in small areas of w and s Mexico and Guatemala, and uncommon to fairly common residents on both slopes from s Nicaragua to Panama.

HABITAT Occurs in tropical and subtropical forests.

BEHAVIOR These still hunters pursue prey inside forests from perches. They make short stoops from perches, but also tail-chase escaping prey. They prey on mammals, birds, large lizards, and snakes.

They soar and glide with wings held level. Powered flight is with strong, quick wing beats. Display flights include high-altitude mutual soaring, undulating flight, and wing fluttering, often accompanied by vocalizations. They soar frequently, most often in late mornings.

The build large stick nests high in tall trees in forests.

ABOVE: Ornate Hawk-Eagle. Adult. Adults have barred underparts. All show heavily banded remiges on somewhat paddle-shaped underwings, and light and dark brown tail bands. Belize. December 2005.
RYAN PHILLIPS—BRRI

LEFT: Ornate Hawk-Eagle. Juvenile. Juveniles are overall whitish, but with a buffy wash on head and breast, dark markings on the flanks, and dark barring on the leg feathers. The long crest is mostly white. Belize. August 2006.
RYAN PHILLIPS—BRRI

MOLT Not studied. Presumably they do not molt all of their remiges annually, resulting in primary wave molt. Apparently all rectrices are replaced annually. Juveniles begin molt into Basic II (second) plumage when about nine months of age.

DESCRIPTION Typical hawk-eagle, with long crest of two feathers. Sexes are alike in plumage, with females larger. Juvenile plumage differs, but Basic II (second) plumage is adult-like. Lores are pale gray, and cere is yellow, paler on juveniles. **Adult.** Striking. Head shows black crown and long crest, orange cheeks and nape, narrow malar stripes, and white throat. Orange breast shows white extending from throat and black line extending from malars. Belly and leg feathers are boldly barred black and white, and white undertail coverts have black tips. Female averages more heavily marked. White underwing coverts have some small black spots. Upperparts are blackish, with some faint white feather edges, becoming more grayish on uppersides of remiges. Long tail above shows blackish and gray-brown bands of equal width and below appears whitish with narrow black bands. Black uppertail coverts show narrow white tips. Eyes are orange-yellow to red-orange. Toes are yellow. **Juvenile.** White head shows long brownish crest and a few brown spots, and markings on white underparts restricted to narrow bold black barring on flanks, leg feathers, and axillars. Upperparts are dark brown, usually showing some narrow pale feather edging. Long tail shows dark brown and pale brown bands of equal width above and appears whitish with narrow black bands below. Eyes change from pale gray to pale yellow. Toes are pale yellow. **Basic II.** Similar to adult plumage, but with whitish areas in head, mix of black and brown feather on upperparts, and uneven barring on underparts. Tail as in adult.

UNUSUAL PLUMAGES No unusual plumages have been described.

HYBRIDS No hybrids have been reported.

ETYMOLOGY *Spizaetus* is from the Greek *spizias*, "hawk," and *aetos*, "eagle." The species name, *ornatus*, is from the Latin *ornare*, "ornate" or "adorned," for its colorful adult plumage.

BLACK AND WHITE EAGLE PLATE 25

Spizaetus melanoleucus

ÁGUILA BLANQUINEGRA

LENGTH 50–64 cm   **WINGSPAN** 110–135 cm   **WEIGHT** 750–1,200 g

IDENTIFICATION SUMMARY Small aerial eagles with feathered tarsi that are most often seen in flight, when they appear buteonine-like. White head, underparts, and underwings strongly contrast with blackish upperparts. Black crown patch, orange cere and gape, black lores and eye-rings, and white leading edge of inner wings are distinctive. Adult and juvenile plumages are very similar. Wingtips fall short of tip of long tail on perched eagles.

TAXONOMY AND GEOGRAPHIC VARIATION Monotypic. No geographical variation.

SIMILAR SPECIES (1) **Ornate Hawk-Eagle** (plate 25) juveniles are similar. But they show dark brown upperparts; narrow bold black barring on axillars, flanks, and leg feathers; and boldly banded undersides of remiges; they also lack black crown patch, lores, and eye-rings and orange ceres and gapes. Their wings are more paddle-shaped.

(2) **Short-tailed Hawk** (plate 18) light-morph hawks are very similar in flight. But they are much smaller, show dark cheeks (especially adults), secondaries darker than primaries, and bare tarsi, and lack bright orange ceres and gapes and black lores and eye-rings.

(3) **Gray-headed Kite** (plate 4) paler juveniles are similar; see under that species for distinctions.

STATUS AND DISTRIBUTION Rare to uncommon and local residents on the Caribbean slope in lowland and foothill areas north to sw Tamaulipas, including most of the Yucatán Peninsula, and south to n Nicaragua, and on both slopes of Costa Rica and onto the Pacific slope of e Panama.

HABITAT Occurs over a variety of habitats, preferring edges of tropical rain forest.

BEHAVIOR These aerial hunters prefer to hunt over more open areas, but also still hunt from concealed and open perches. They capture a wide variety of prey, including mammals, birds, reptiles, and amphibians, usually after a spectacular stoop from the air, but they also tail-chase prey through forests.

They soar and glide with wings held level. Powered flight is with strong, quick wing beats. Display flights include mutual soaring and undulating flight, often accompanied by vocalizations.

Large stick nests are built high in tall trees in forests.

MOLT Not studied. Presumably they do not molt all of their remiges annually, resulting in primary wave molt. Apparently all remiges are

OPPOSITE PAGE: Black and White Eagle. Adult. Overall black and white, with bright orange cere and gape, black beak, and black around eyes. Belize. August 2009. YERAY SEMINARIO

LEFT: Black and White Eagle Adult. Overall white below, with faint markings on flight feathers. Wing shape is more like those of *Buteos*. Belize. August 2006. LIOR KISLEV

ABOVE: Black and White Eagle. Juvenile. Juveniles are much like adults, but show pale feather edges on upperparts and more and narrower tail bands. Belize. August 2009. YERAY SEMINARIO

replaced annually. Juveniles begin molt into adult plumage when about nine months of age.

DESCRIPTION Sexes are alike in plumage, with females noticeably larger. Legs are completely feathered. Beak is black, and toes are yellow to orange-yellow. Wing shape of flying eagles is more buteonine and not paddle-shaped. **Adult.** White head shows small black crown patch, including short crest, bright orange cere and gape, black lores and eye-rings, and yellow to orange-yellow eyes. Underparts, underwing coverts, and leg feathers are white. Undersides of white remiges and rectrices show several narrow black bands, the subterminal wider, and the black tips of the outer primaries have white banding. Upperparts are black, and uppertail shows black and pale gray bands of equal width. **Juvenile.** Very similar in plumage to adults and quite difficult to separate in the field. Black crown patch and upperwing coverts show faint pale spotting. Upperparts have a brownish cast. Undersides of white remiges usually show more narrow bands compared to adults. Black tail bands are somewhat narrower than those of adults. Eyes and cere and gape are paler than those of adults.

FINE POINTS See Phillips and Seminario (2009) for more details of ageing.

UNUSUAL PLUMAGES No unusual plumages have been described.

HYBRIDS No hybrids have been reported.

ETYMOLOGY *Spizaetus* is from the Greek *spizias*, "hawk," and *aetos*, "eagle." The species name, *melanoleucus*, is from the Greek *melas* and *leukos*, "black" and "white." Often referred to as Black-and-white Hawk-Eagle; they do not hunt in forests as do hawk-eagles but are aerial hunters and therefore not hawk-eagles.

REFERENCES Phillips and Seminario 2009.

LAUGHING FALCON PLATE 27
Herpetotheres cachinnans

HALCÓN GUACO

LENGTH 44–52 cm **WINGSPAN** 75–90 cm **WEIGHT** 520–770 g

IDENTIFICATION SUMMARY Distinctive, stocky, medium-sized fal-conids that have large pale heads with extensive dark masks. Plumages of adults and juveniles are almost alike. Wings are paddle-shaped and show pale primary panels, and wingtips barely extend beyond folded secondaries on perched falcons. Long dark tail shows numerous narrow pale bands.

TAXONOMY AND GEOGRAPHIC VARIATION Nominate subspecies occurs throughout. Alleged subspecies *H. c. chapmani* does not differ much from nominate.

SIMILAR SPECIES No other raptor is similar.

STATUS AND DISTRIBUTION Uncommon to fairly common breeding residents on both slopes from c Mexico, including the Yucatán Penin-sula, to Panama.

HABITAT Occurs in a variety of open wooded habitats, including for-est edge and clearings, gallery forest, savannah, second growth, and plantations. Apparently absent from continuous humid forest.

Laughing Falcon. Adult. All plumages are distinctive, with large head showing buffy crown, dark mask across nape, and wide buffy hind collar. Long tail shows two or three pale bands. Venezuela. January 2006. wsc

BEHAVIOR They prey primarily on snakes, which they hunt from elevated perches, executing short stoops to grab them. They also take occasional lizards and small rodents.

Active flight is direct with rapid wing beats, with interspersed glides on level wings. They do not soar.

They usually nest in tree cavities, but also in old stick nests of other species, in holes in cliffs, and on epiphytes.

Distinctive vocalizations can be heard from long distances.

MOLT Not studied. Presumably molt is completed annually. Juveniles molt directly into adult plumage.

DESCRIPTION Sexes are alike in plumage, and juvenile plumage is almost like adult's. Eyes are dark brown. **Adult.** Creamy head has bold black face mask that extends around nape, and neck is creamy. Upperparts are dark brown, and creamy underparts are unmarked. Short, paddle-shaped wings are dark brown above with creamy primary panels and creamy below with narrow dark barring remiges. Long, graduated blackish tail shows three pale bands. Some show faint dark markings on underwing coverts. Some adults in South America show narrow pale scapulars. Cere is yellow and legs are dull dirty yellow. **Juvenile.** Almost like adult, but with faint narrow crown streaking, narrow rufous feather edges on upperparts, faint narrow streaks on underparts, narrower dark terminal band on remiges, and buffy tail bands. Cere and legs are dull yellowish.

UNUSUAL PLUMAGES No unusual plumages have been described.

HYBRIDS No hybrids have been reported.

ETYMOLOGY *Herpetotheres* comes from the Greek *herpeton*, "reptile," and *theras*, "hunter." The species name, *cachinnans*, comes from the Latin *cachinnans*, "laughing aloud."

BARRED FOREST-FALCON PLATE 28

Micrastur ruficollis

HALCÓN MONTÉS BARREADO

LENGTH 31–40 cm   **WINGSPAN** 46–60 cm   **WEIGHT** 130–250 g

IDENTIFICATION SUMMARY Smallest of three secretive rain forest raptors. They are more often heard than seen. All show dark upperparts, a variable amount of dark barring on underparts, short paddle-shaped wings, and long dark tail with narrow whitish bands and graduated tip. Adult and juvenile plumages differ, with buffy and white morph in the latter. Perched forest-falcons show short primary projection, barely beyond secondaries.

TAXONOMY AND GEOGRAPHIC VARIATION The subspecies *M. r. interstes* occurs throughout, with no geographic variation. The named subspecies *M. r. guerilla* from Nicaragua north is not valid, as the minor differences are clinal.

Barred Forest-Falcon. Adult. Adults have gray heads and narrowly barred white underparts. Belize. October 2006.

RYAN PHILLIPS—BRRI

SIMILAR SPECIES (1) **Tiny Hawk** (plate 10) is similar in having dark upperparts and barred underparts, but they have shorter legs, longer primary projection, and wide pale tail bands, and lack bright eye-ring and face skin.

(2) **Slaty-backed Forest-Falcon** (plate 28) is somewhat similar, but they are larger, have unbarred underparts, and show narrow dark line in pale uppertail bands.

(3) **Double-toothed Kite** (plate 6) adults are similar in having dark upperparts and barred underparts, but have shorter legs, longer primary projection, and wide pale tail bands, and lack bright eye-ring and face skin.

Barred Forest-Falcon. Juvenile. Juveniles have brown heads and narrow rufous barring on buffy underparts. Note pale face ring and long graduated tail. Ecuador. June 2010. NICK ATHANAS

STATUS AND DISTRIBUTION Uncommon to common, but local and unobtrusive on lowlands and lower slopes of the Pacific slope locally in s Mexico and from c Guatemala to n Costa Rica, and locally in c Costa Rica and on the Caribbean slope from c Mexico (but not in n Yucatán Peninsula) to the south through Panama.

HABITAT Occurs in lower and mid-levels of lowland primary rain forest; less common in second growth.

BEHAVIOR These still hunters of the forest understory are most active at dawn and dusk. They also actively follow army ant swarms, preying on both insects and birds that follow the ants. They prey on a variety of small birds, lizards, small mammals, large insects, bats, and frogs, and there is at least one report of them eating fruit. They often show an owl-like facial ring and no doubt locate some prey by sound.

Powered flight is with quick wing beats. The glide with wings held level, but do not soar.

They lay eggs and raise chicks in tree cavities.

They are quite vocal and are often heard, mostly at dawn, but also at dusk, and less often throughout the day. The call most often heard is a yapping similar to the bark of a small dog.

MOLT Not studied. Presumably molt is completed annually. Juveniles molt directly into adult plumage.

DESCRIPTION Adults are similar in plumage, with females larger. Juvenile plumage is different and dimorphic. **Adult.** Adults have gray head and upperparts, white underparts with narrow gray barring (finer on males), and a long black tail with three narrow white bands. White undersides of flight feathers are heavily barred, and white underwing coverts are finely barred gray. White undertail coverts are finely barred gray, sometimes lacking on males. Long legs are orange-yellow. Cere, lores, and eye-rings are bright orange, and eyes are pale brown to pale hazel. **White-morph juvenile.** Head and upperparts are brownish-black, and whitish underparts show a variable amount of gray barring (on some less heavily marked), and long black tail with three or four (even five or six) narrow white bands. They show a whitish hind collar. Long legs are dull yellow to yellow. Cere, lores, and eye-rings are dull yellow-green. Eyes are dark brown. **Buffy-morph juvenile.** Some juveniles show a buffy wash on underparts and hind collar. **Rufous-morph juvenile.** Some juveniles have rufous heads, upperparts, and a wide breast band and the rest of the underparts are narrowly barred.

UNUSUAL PLUMAGES No unusual plumages have been described.

HYBRIDS No hybrids have been reported.

ETYMOLOGY *Micrastur* comes from the Greek *mikroa*, "small," and the Latin *astur*, "hawk." The species name, *ruficollis*, is from the Latin *rufus*, "red," and *collaris*, "collared" or "of the neck," for the rufous collar shown by some juveniles.

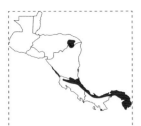

SLATY-BACKED FOREST-FALCON PLATE 28

Micrastur mirandollei

HALCÓN MONTÉS LOMO PIZARREÑO

LENGTH 40–44 cm **WINGSPAN** 65–71 cm **WEIGHT** 420–560 g

IDENTIFICATION SUMMARY Medium-sized, secretive rain forest raptors restricted to e Panama and Caribbean slope of w Panama and Costa Rica. They are more often heard than seen. All show dark upperparts, white underparts, short paddle-shaped wings, and a long, dark, graduated tail with narrow whitish bands. Adult and juvenile plumages differ, with marked underparts on the latter. Perched forest-falcons show short primary projection, barely beyond secondaries.

TAXONOMY AND GEOGRAPHIC VARIATION Monotypic, with no geographic variation.

SIMILAR SPECIES (1) **Barred Forest-Falcon** (plate 28) is somewhat similar; see under that species for distinctions.

(2) **Collared Forest-Falcon** (plate 28) light-morph adult is similar, but they are larger and even longer-tailed and show distinctive nuchal collar. Light-morph juvenile has heavily barred underparts.

(3) **Bicolored Hawk** (plate 11) juveniles are similar. But they have a wide pale collar on hind neck, creamy underparts with buffy leg feathers, and wider pale tail bands.

(4) **Semiplumbeous Hawk** (plate 12) is similar; see under that species for distinctions.

STATUS AND DISTRIBUTION Uncommon to rare but local resident breeders in rain forest of e Panama and on the Caribbean slope of w Panama and Costa Rica, and locally on the Pacific coast of Nicaragua. There are recent valid records from se Honduras and adjacent Nicaragua.

HABITAT Occurs in humid lowland rain forest.

BEHAVIOR These still hunters of the interior of rain forests are most active at dawn and dusk. They apparently prey on a variety of small birds, as well as lizards and small mammals. They often show an owl-like facial ring and no doubt locate some prey by sound. While they are most commonly still hunters, they have been reported foraging on the forest floor. A report of them mimicking bird calls to attract them in close is highly doubtful.

Powered flight is with quick wing beats. The glide with wings held level, but do not soar.

Their nests have not been described. Alleged description from Amazonia was of an accipiter.

MOLT Not studied. Presumably molt is completed annually. Juveniles molt directly into adult plumage.

DESCRIPTION Adults are alike in plumage, and juvenile plumage is similar. Beak is black. **Adult.** Adults have slate-gray head and upperparts, including upperwing coverts; white unmarked throat and underparts; and long, black graduated tail with three pale gray bands with a narrow darker band in the center of each. Undersides of white flight feathers are heavily barred, more whitish and less heavily barred on bases, and white underwing coverts are unmarked. Eyes are dark brown. Cere, lores, and eye-rings are bright yellow. **Juvenile.** Juveniles are similar to adults, but have brownish cast to head and upperparts and brownish mottling on white underparts, heavier on breast. Cere, lores, and eye-rings are bright yellow, bleeding onto the base of the otherwise dark beak.

UNUSUAL PLUMAGES No unusual plumages have been described.

HYBRIDS No hybrids have been reported.

ETYMOLOGY *Micrastur* comes from the Greek *mikroa*, "small," and the Latin *astur*, "hawk." The species name, *mirandollei*, is after a resident of Suriname, M. Mirandolle.

COLLARED FOREST-FALCON

PLATE 28

Micrastur semitorquatus

HALCÓN MONTÉS ACOLLARADO

LENGTH 50–59 cm   **WINGSPAN** 75–87 cm   **WEIGHT** 465–820 g

IDENTIFICATION SUMMARY The largest *Micrastur*, long and lanky and with a relatively longer graduated tail, compared to other forest-falcons. They occur in three color morphs: light (common), buffy (less common), and dark (uncommon). Light and buffy morphs show distinctive dark semi-collar and pale hind collar. All show paddle-shaped wings and a long, graduated dark tail with narrow whitish bands. When perched, they show short primary projection, barely beyond secondaries. They are more often heard than seen.

TAXONOMY AND GEOGRAPHIC VARIATION Monotypic, with no geographic variation.

SIMILAR SPECIES (1) **Slaty-backed Forest-Falcon** (plate 28) adults are somewhat similar; see under that species for distinctions. (2) **Bicolored Hawk** (plate 11) juveniles are similar; see under that species for distinctions. (3) **Gray-headed Kite** (plate 4) light juvenile is somewhat similar to adult light Collared Forest-Falcon; see under that species for distinctions. (4) **Hook-billed Kite** (plate 4) juveniles perched also show white hind collar, but they are much smaller, with longer primary projection, barred underparts, and shorter paler tail with dark bands.

STATUS AND DISTRIBUTION Uncommon to fairly common breeding residents in lowland and foothills of both slopes from c Tamaulipas (and Nueva Leon) and s Sinaloa to Panama, but absent on the interior plateau of Mexico and highlands from s Mexico to Nicaragua.

HABITAT Occurs in more open habitats compared to the other two *Micrasturs*, including edge of forest, along rivers and streams, gallery forest, second growth, and plantations.

BEHAVIOR These forest predators hunt from perches as well as pursue prey through forest on the wing or on foot or along branches with surprising agility. They prey to a large extent on mammals and medium-sized to quite large birds, as well as lizards and snakes. They are most active at dawn and dusk. They often show an owl-like facial ring and no doubt locate some prey by sound. A report of them mimicking bird calls to attract them in close is highly doubtful.

Powered flight is with rapid wing beats, rarely leaving forests, but occasionally above the canopy or across an open area. They glide with wings held level, but do not soar.

Their nests are in tree cavities.

OPPOSITE PAGE: Collared Forest-Falcon. Adult light. Light adults have white underparts and hind collar. Dark slash beneath white cheek is distinctive. All have long graduated tails. Mexico. April 2008.
ANTONIO HIDALGO

They are quite vocal and are usually seen more often than heard. Usually call is a single drawn-out note, but can also be a repeated set of notes. They call at any time of day, but usually more commonly at dawn.

MOLT Not studied. Presumably molt is completed annually. Juveniles molt directly into adult plumage.

DESCRIPTION Adults are alike in plumage, and juvenile plumage is similar. They occur in three color morphs: light, buffy, and dark. Beak is black. Eyes are dark brown, and cere, lores, and eye-rings are dull greenish-yellow. **Adult light.** White head shows black crown and semi-collar, and white hind neck forms a nuchal collar. Upperparts are blackish, and underparts are white. Underwings show whitish flight feathers heavily barred, more whitish and less heavily barred on bases, and unmarked white underwing coverts. Long, blackish tail is graduated and shows four or five narrow white bands. Long legs are yellow to orange-yellow. **Adult buffy.** Similar to light adult, but with buffy replacing white in the plumage. Some adults of any morph can show a narrow pale superciliary line. **Adult dark.** Entire body and wing coverts are slate black, and black undertail coverts show narrow white barring; lack pale cheeks and hind collar. Tail and flight feathers are like those of light adult. **Juvenile light.** Color and pattern are similar to light adult, but upperparts are browner with some rufous feather edging, underparts are boldly barred dark gray, and tail has one or two additional white bands. Cheek patch and hind collar are less obvious. Legs are dull yellow. **Juvenile buffy.** Color and pattern are similar to buffy adult, but upperparts are browner with extensive rufous feather edging, underparts are marked with dark brown spotting on breast and barring on flanks, and tail has one or two additional white bands. Legs are dull yellow. **Juvenile dark.** Color and pattern are similar to dark adult, but upperparts are browner with some rufous feather edging, belly is narrowly barred rufous, and tail has one or two additional white bands. Legs are dull yellow.

UNUSUAL PLUMAGES No unusual plumages have been described.

HYBRIDS No hybrids have been reported.

ETYMOLOGY *Micrastur* comes from the Greek *mikroa*, "small," and the Latin *astur*, "hawk." The species name, *semitorquatus*, is from the Latin *semi*, "half," and *torquatus*, "collared."

RED-THROATED CARACARA

PLATE 26

Ibycter americanus

CHUPACACAO VIENTRE BLANCO

LENGTH 43–56 cm   **WINGSPAN** 96–124 cm   **WEIGHT** 510–710 g

IDENTIFICATION SUMMARY Distinctive black and white raptors, with a large red throat patch (adults). Juvenile plumage is almost like adults. They are social and communal and specialize on eating bee and wasp nests. Wingtips do not reach the tail tip on perched caracaras.

TAXONOMY AND GEOGRAPHIC VARIATION Monotypic, with no geographic variation.

SIMILAR SPECIES No other raptors in our area are similar.

STATUS AND DISTRIBUTION Rare and very local in lowland rain forest in n Honduras and sw Costa Rica and n Panama; more extensive in e Panama. They were formerly more widespread and common.

HABITAT Occurs in a variety of lowland forest habitats.

BEHAVIOR They specialize in preying on the nests and adults of wasps and bees, but also take caterpillars and turtle eggs, and even some fruit and soft seeds. They have been reported to fly at these nests and knock them to the ground, and to hang on the nest and break into it with the beak. They are quite social and occur in small groups, which often forage together and are quite noisy.

Powered flight is with slow, somewhat labored wing beats, rather corvid-like. They glide with wings held level and never soar.

Their nesting has not been studied in our area.

MOLT Not studied. Presumably completed annually. Juveniles molt directly into adult plumage.

DESCRIPTION Sexes are alike in plumage, and juvenile plumage is almost like adult's. **Adult.** Overall glossy black, except for white

Red-throated Caracara. Adult. Adults are overall blackish but with colorful head and throat and white belly. French Guinea. December 2012. SEAN MCCANN

Red-throated Caracara.
Adult. Adults are overall
blackish but with colorful
face and beak. French
Guinea. December
2012. SEAN MCCANN

belly, leg feathers, and undertail coverts, and large bare red throat patch. Face skin, large throat patch, and legs are reddish, cere and base of mandible are pale blue-gray, eyes are red to reddish-brown, and beak is bright yellow. Long black tail has a rounded tip. **Juvenile.** Much like adult, but black plumage is less glossy, face skin is grayish with some red highlights on the throat, eyes are brown, beak is pale yellow, cere is yellow, and legs are paler red.

FINE POINTS Juveniles and adults have some paler gray feathers bordering the bare facial and throat skin.

UNUSUAL PLUMAGES No unusual plumages have been described.

HYBRIDS No hybrids have been reported.

ETYMOLOGY *Ibycter* comes from the Greek *ibukter*, "shouter." The species name, *americanus*, is from America.

Red-throated Caracara.
Juvenile. Juveniles are
like adults but plumage
is less glossy and face
pattern is duller. Costa
Rica. June 2013.
CHRIS JIMÉNEZ

CRESTED CARACARA PLATE 26

Caracara plancus

CARANCHO

LENGTH 54–60 cm **WINGSPAN** 118–132 cm **WEIGHT** 800–1,300 g

IDENTIFICATION SUMMARY Large, unusual falconids. Their bold black and white plumage, orange to pink to yellow face, large horn-colored beak, large crested head, and long neck in flight are distinctive. All show large pale primary panels and pale tail with wide dark terminal band. The long head and neck of flying caracaras is distinctive. They have three plumages: adult, Basic II (second plumage), and juvenile. Wingtips fall just short of the tail tip on perched caracaras.

TAXONOMY AND GEOGRAPHIC VARIATION The subspecies *C. p. cheriway* occurs throughout our region, with no geographic variation. Some authorities recognize the Southern Caracara as a separate species and call our bird the Northern Caracara (*Caracara cheriway*), but the differences are minimal and are not at the species level.

Crested Caracara. Adult. Adults are overall black and white with large orange face skin, pale beak, and orange legs. Texas USA. June 2015. WSC

Crested Caracara. Adult. Adults' long wings show pale primary panels, and long tail has wide dark subterminal band. Texas USA. October 2013. PATTY RAINEY

SIMILAR SPECIES (1) **Common Black Hawk** (plate 16) juvenile has similar wing and tail pattern, but lacks belly band and long neck.

STATUS AND DISTRIBUTION Uncommon to common breeding residents in open areas from the coastal lowlands on both slopes from n Mexico (but not most of the Yucatán Peninsula) to Panama.

HABITAT Occurs in a variety of open and semi-open habitats, usually in lowlands, but not in rain forest.

BEHAVIOR They are primarily scavengers, but also pirate prey from other birds, especially vultures, forcing them to give up or disgorge food. They also prey on live birds, small mammals, reptiles and amphibians, and insects. They often harass other raptors and dominate Black and Turkey Vultures at carcasses. Caracaras spend considerable time on the ground foraging for food and cruise highways searching for road-killed animals. Allopreening by the adults has been observed, including with Black and Turkey Vultures.

Active flight is with medium-slow, steady, almost mechanical wing beats. They soar with wings held level, with leading and trailing edges straight. Caracaras appear eagle-like when soaring because of their long neck and straight wings. They glide with wings bowed, with wrists cocked forward and above the body, and wingtips pointed down. When pursuing other raptors on the wing, they are agile and acrobatic.

When excited, they throw the head back and snap it forward or roll the back of the head across the shoulders, at the same time giving their rattle call. They form communal night roosts, especially in winter.

Unlike falcons of the genus *Falco*, they build their own nest, a large bulky stick nest in a small tree, bush, cactus, or yucca. Vocalizations include a low rattle and a single "wuck" note.

MOLT Annual molts are complete. Juveniles begin molt into Basic II (second) plumage in their first spring.

DESCRIPTION Large, unusual falconids, with a large blue-gray to horn-colored beak. Sexes are alike in plumage, but females are slightly larger. Eyes are medium brown. **Adult.** Head and neck are creamy, except for black crown and long crest and large yellow to orange cere and facial skin. Upper back is black with fine white barring; lower back and upperwing coverts are dark blackish-brown. Long neck and upper breast are creamy, with black barring in mid-breast. Black belly forms solid band, and leg feathers are black. Wings are black except that mostly white outer primaries form large wing panels. Underwing coverts are black. Uppertail and undertail coverts are white. White tail has many narrow black bands and a wide dark terminal band. Long legs are yellow to yellow-orange. **Juvenile.** Similar to adult except back and crown are dark brown, neck and throat are buffy, and mid-breast and upper back are streaked with brown rather than barred with black. In fresh plumage, upperparts show pale feather edges. Facial skin is gray to pink, eyes are dark brown, and legs are gray, becoming pale yellowish-gray. **Basic II (second plumage).** Almost like adult, but overall more brownish-black. Brownish-black crown shows narrow rufous streaking. Barring on breast and upper back is less well defined. Face skin is orange.

271

Brownish-black belly can show pale barring. Long legs are dull yellow gradually becoming yellow.

FINE POINTS They sometimes continue to flap when soaring. Facial skin changes color depending on mood, ranging from orange on normal adult to bright yellow, indicating excitement.

UNUSUAL PLUMAGES Partial albinos have been reported from w Mexico. An amelanistic specimen was collected in Argentina.

HYBRIDS No hybrids have been reported.

ETYMOLOGY "Caracara" and *Caracara* come from the Tupi native American (Brazil) onomatopoeic name. The species name, *plancus*, comes from the Latin *plancus*, "a kind of eagle." **Note.** Audubon's Caracara is another name for our race.

YELLOW-HEADED CARACARA

PLATE 27

Milvago chimachima

CHIMACHIMA

LENGTH 40–47 cm   **WINGSPAN** 81–95 cm   **WEIGHT** 270–400 g

IDENTIFICATION SUMMARY Small, slender, long-tailed caracara. Adult and juvenile plumages differ, but both show buffy primary panels. Adults are creamy and dark brown, and juveniles are heavily streaked. Wingtips fall somewhat short of the tail tip on perched caracaras.

TAXONOMY AND GEOGRAPHIC VARIATION Monotypic, with no geographic variation. Four subspecies have been described, but differences are minor, with much overlap among them.

SIMILAR SPECIES No other raptors are similar.

STATUS AND DISTRIBUTION Uncommon to fairly common on the Pacific slope from n Costa Rica through Panama and on both slopes in e Panama, with recent (vagrant?) records from s Honduras.

HABITAT Occurs in a variety of open areas, including savannahs, agriculture, and pastures, usually not far from water.

Yellow-headed Caracara. Adult and juvenile. Adults (right) are overall buffy below with narrow dark eye-lines. Juveniles (left) appear darker and are heavily streaked below. Brazil. October 2006. NICK ATHANAS

Yellow-headed Caracara. Adult. Adults are buffy below and dark brown above. Note narrow dark eye-line, paddle-shaped wings with pale primary panel, and long tail. Colombia. September 2012.

STU ELSOM

BEHAVIOR Primarily scavengers, especially feeding on roadkills, but they also eat insects, crabs, frogs, and toads, as well as some vegetable matter. They regularly associate with cattle, walking among them and running after prey displaced by the cows. They spend much time flying low, patrolling for food, especially along roads.

Powered flight is with floppy, somewhat weak wing beats. They glide with wings held level or slightly cupped. They do not soar.

MOLT Not studied. Presumably molt is completed annually. Juveniles molt directly into adult plumage.

DESCRIPTION Small, slender caracara, with a small weak beak. Sexes are alike in plumage, with females barely larger than males. **Adult.** Creamy head shows narrow dark eye-lines that extend toward the nape. Underparts and underwing coverts are unmarked creamy-buff. Back and upperwing coverts are dark brown. Primaries have wide dark wingtips and creamy bases forming pale wing panels. Dark brown secondaries show faint pale banding on undersides. Uppertail coverts are creamy. Long creamy tail shows some narrow dark bands and a wide dark subterminal band. Eyes are medium to dark brown.

Beak is pale blue. Cere and eye-ring are orange-yellow to pinkish. Legs are dull yellow. **Juvenile.** Overall dark and streaky. Head shows streaked crown and throat and wide dark eye-line. Dark brown upperparts show some faint pale feather edges. Creamy underparts are heavily streaked dark brown. Primaries have wide dark wingtips and creamy bases forming pale wing panels. Dark brown secondaries show pale banding on undersides. Uppertail coverts are creamy. Long tail shows narrow dark brown and cream banding. Eyes are dark brown. Beak is yellowish, and cere and eye-ring are pale yellow to pinkish. Legs are dull yellow.

UNUSUAL PLUMAGES No unusual plumages have been described.

HYBRIDS No hybrids have been reported.

ETYMOLOGY *Milvago* is from the Latin *milvus*, "kite," and *ago*, "resembling." The species name, *chimachima*, is the Native American (Argentina) name for this species.

Yellow-headed Caracara. Juvenile. Juveniles are heavily streaked below and dark brown above. Note narrow dark eye-line, paddle-shaped wings with pale primary panel, and long heavily banded tail. French Guinea. June 2010. MICHEL GIRAUD-AUDINE

AMERICAN KESTREL PLATE 29

Falco sparverius

CERNÍCALO AMERICANO

LENGTH 22–26 cm   **WINGSPREAD** 52–61 cm   **WEIGHT** 97–140 g

IDENTIFICATION SUMMARY Our smallest and most colorful falcon. Gray head with rufous crown patches and white cheeks with two black mustache marks (vertical black bars) are distinctive. Males and females have different plumages. This is our only falcon that hovers and kites regularly. Wingtips don't reach tail tip on perched kestrels.

TAXONOMY AND GEOGRAPHIC VARIATION *F. s. sparverius* occurs as a migrant and winter visitor throughout. Three resident races differ little: *F. s .peninsularis* in Baja California and w Mexico, *F. s. tropicalis* in s Mexico to n Honduras, and *F. s. nicaraguensis* from Honduras to Costa Rica. South American race *F. s. isabellinus* is smaller, has little or no spotting on underparts, and is generally a richer color and occurs in Panama and s Costa Rica.

SIMILAR SPECIES (1) **Merlin** (plate 29) is darker, lacks the two mustache marks, and is larger (female) or chunkier (male). In flight, Merlins show darker underwings, broader wings, and larger head than American Kestrels.

American Kestrel. Adult male. Adult males have rufous backs and blue-gray wing coverts. All show colorful head with two black mustache marks. Texas USA. December 2013. WSC

American Kestrel. Adult female. Adult females have buffy streaked underparts and brown tail with dark brown bands, the subterminal band wider. California USA. February 2008.
RYAN PHILLIPS—BRRI

(2) **White-tailed Kite** (plate 5) is larger and overall white and gray, hovers more vertically like a kingfisher, and has an unbanded white tail.

(3) **Pearl Kite** (plate 6) appears similar when perched; see under that species for distinctions.

(4) **Peregrine** (plate 31) is much larger and darker and has longer, broader wings and a much larger head.

STATUS AND DISTRIBUTION Uncommon to fairly common breeding residents in Baja California, n and c areas of Mexico and higher mountains from s Mexico to Nicaragua. They are also rare breeding residents in open areas of e Panama and are much more widespread in winter throughout.

HABITAT Occurs in a variety of open areas, with residents occurring along edges of forests and winter visitors occurring in open agricultural areas.

BEHAVIOR Usually seen hovering or sitting on exposed perches, such as poles, wires, or treetops, where they hunt for rodents,

American Kestrel. Juvenile male. Juvenile males have dark streaking on their breasts. California USA. July 1994. wsc

insects, birds, lizards, or snakes. In a character that is unique to kestrels, they wag the tail up and down when excited.

Active flight is light and buoyant; however, American Kestrels sometimes chase birds in a direct, rapid, Merlin-like fashion. They soar on flat wings, often with the tail fanned. They glide on level wings or with wrists lower than the body and wingtips curved upward. This is the only American falcon to hunt regularly by hovering (wings flapping) or, in stronger winds, by kiting (wings held steady).

They nest in tree cavities, but will readily use holes in cliffs and crevices in barns and buildings as well as nest boxes. They can be vociferous, and their easily recognized "killy-killy" call carries some distance.

MOLT Annual molt of adults is complete. Juveniles undergo a more or less complete pre-formative body molt in late summer.

DESCRIPTION Sexes have different adult and juvenile plumages. In flight, underwings appear pale. Juvenile plumages are similar to those of same sex adults. Females are slightly larger, but there is considerable overlap in size. Eyes are dark brown. Cere, eye-ring, and leg colors are orange to yellow, paler on juveniles. **Adult male.** Blue-gray crown has a variably sized rufous crown patch that is sometimes lacking. White cheeks show two black mustache marks, one below the eye and the other at the rear of the auriculars. Rufous back is barred black on lower half. Blue-gray upperwing coverts show small black spots. Breast color varies from whitish to deep rufous, and belly is white with black spots, heavier on flanks. Black primaries have a row of white circles on the trailing edge, visible on underwings when wing is backlit. Leg feathers are white to rufous. Typical tail is rufous with wide black subterminal band and with terminal band that is white, rufous, or gray, or some combination of those colors. Outer tail feathers are white with black bands. Tail patterns vary considerably, including some with little or no rufous. Residents in Panama and s Costa Rica (*isabellinus*) are smaller and brighter rufous, have fewer dark bars on the back, and lack black spots on the flanks. **Adult female.** Head is like that of adult male but paler. Back and upperwing coverts are reddish-brown with dark brown barring. Pale underwings have a row of buffy circles on trailing edge of backlit wing, less noticeable than those on wing of male. Creamy underparts are heavily streaked with reddish-brown. Leg feathers are clear white to creamy. Reddish-brown tail has eight or more narrow dark brown bands and a noticeably wider dark subterminal band. Residents in Panama and s Costa Rica (*isabellinus*) are smaller and have indistinct streaking on their underparts. **Juvenile male.** Similar to adult male, but whitish breast is heavily streaked with black, back is completely barred, and crown patch has black shaft streaks. They usually molt into adult body plumage a few months after fledging, but some retain juvenile characteristics for almost a year. **Juvenile female.** Like adult female, but often with dark subterminal tail band the same width as or slightly wider than the other dark bands.

FINE POINTS American Kestrels possess a pair of false eyes, or ocelli, on the nape. This is thought to be protective coloration in that the watching "eyes" will deter potential predators.

UNUSUAL PLUMAGES Partial albinos with some to almost all white feathers have been reported. Specimens that are partially melanistic are mostly dark. Specimens of gynandromorphs (individuals exhibiting characteristics of both genders) were sexed internally as female, but with many male feathers.

HYBRIDS No hybrids have been reported.

ETYMOLOGY *Falco* is from the Latin *falx*, "sickle," in reference to the falcon's wing shape in flight or, according to another source, to the shape of their beak and talons. The species name, *sparverius*, in Latin means "pertaining to a sparrow," after the falcon's first common name, Sparrow Hawk. It had been misnamed after the European *Accipiter nisus*, the Sparrowhawk. "American Kestrel" comes from its Eurasian counterpart, the Eurasian Kestrel, *Falco tinnunculus*. "Kestrel" is an old English name for *F. tinnunculus*.

MERLIN PLATE 29
Falco columbarius
ESMEREJÓN

LENGTH 24–30 cm **WINGSPREAD** 53–68 cm **WEIGHT** 129–236 g

IDENTIFICATION SUMMARY Small, dashing, compact falcons. Two races occur: Taiga and Prairie. Both have different adult male and female and juvenile plumages; however, the latter two are difficult to distinguish in the field. All lack the bold mustache mark of other falcons, having at most only a faint one. Dark tail shows three or four narrow pale bands. They do not hover. On perched falcons, wingtips do not reach tail tip.

TAXONOMY AND GEOGRAPHIC VARIATION Taiga Merlin is *F. c. columbarius*, and Prairie Merlin is *F. c. richardsoni*.

SIMILAR SPECIES (1) **American Kestrel** (plate 29) is smaller, has a lighter, more buoyant flight, and in flight shows smaller head and narrower wings with pale undersides. Perched Kestrels have a rufous tail and back and two distinct black mustache (vertical) marks on each side of the face.

(2) **Peregrine Falcon** (plate 31) is larger and has relatively longer wings, a distinct mustache mark, and a larger head. Juvenile male Peale's Peregrine may appear similar to female Black Merlin, but

Merlin. Adult male. Adult males have blue-gray upperparts. All show faint mustache marks. Wisconsin USA. July 1992. WSC

when perched, the Peregrine's wingtips reach or almost reach the tail tip.

(3) **Prairie Falcon** (plate 31) is much larger than Prairie Merlin, has distinct mustache marks, and dark auriculars, and in flight shows dark center of pale underwing.

(4) **Bat Falcon** (plate 30) also flies fast and direct, but has completely dark head, large white area on the throat, and rufous leg feathers.

STATUS AND DISTRIBUTION Taiga Merlins are rare and local migrants and winter visitors throughout our area. Prairie Merlins most likely go no farther than n Mexico.

HABITAT Occurs in variety of open and semi-open areas.

BEHAVIOR These dashing falcons hunt birds on the wing. Their rapid hunting flight is typically in a direct line over open forest (over grasslands for Prairie Merlins). They also hunt from exposed perches, making rapid forays after prey, and regularly harass larger birds, including gulls and other raptors, using speed to surprise a flock and snatch a bird that is slow to react. They fly fast enough to tail-chase swallows, swifts, and shorebirds and are capable of very sudden changes in direction and making spectacular aerial maneuvers. They capture flying insects, especially dragonflies, in flight. They often force small birds to fly up to a great altitude and capture them when they try to drop back to earth. Merlins have an unusual flight mode when attacking a bird from low altitude. They flap their wings in quick bursts interspersed with glides, producing an undulating but rapid flight that appears like that of passerines. This mode of attack may allow them to be mistaken for a passerine until it is too late for the prey to escape.

Active flight is direct and rapid, with strong, quick wing beats. They soar with level wings and with the tail somewhat fanned. They glide on level wings or with wrists lower than the body and the wingtips curved upward.

Their primary vocalization is a rapid, high-pitched "ki ki ki."

MOLT Annual molt is complete. Juveniles begin molt into adult plumage in their first spring.

DESCRIPTION Measurements above are for Taiga Merlins. Prairie Merlins are a bit larger. Sexes are different in adult plumage and size, with females noticeably larger. Juvenile plumage is similar to that of adult female. Eyes are dark brown. Cere and eye-ring colors are greenish-yellow to yellow. Leg color is yellow, but orange-yellow on breeding males. **Taiga Merlins** always show dark underwings. **Taiga adult male.** Head consists of slate-blue crown with fine black streaking, buffy superciliary line, buffy cheek with darker marking behind the eye, a faint dark mustache mark, faint whitish or rufous markings

Merlin. Adult female. Adult females have brown upperparts. Underwings are dark, and dark tail has pale bands. Saskatchewan, Canada. July 1992.
BRIAN SULLIVAN

on the hind neck, and a white, unstreaked throat. Back and upper-wing coverts are slate-blue. Whitish underparts have reddish-brown to dark brown streaking. Breast often has light rufous wash. Whitish leg feathers are lightly streaked and have a rufous wash. Whitish un-dertail coverts have light streaking. Black tail has three slate-blue bands and a wide white terminal band. **Taiga adult female.** Head is like that of adult male, except crown is dark brown with thicker black streaking. Back and upperwing coverts are medium to dark brown, sometimes with a grayish cast, and brown uppertail coverts have a grayish cast. Creamy underparts have heavy dark brown streaking, with barring on the flanks. Creamy leg feathers and undertail cov-erts are lightly streaked. Dark brown tail has four usually incomplete buffy bands and a wide white terminal band. **Taiga juvenile.** Similar to Taiga adult female, but back and upperwing and uppertail coverts are dark brown without a grayish cast. Dark brown tail has four buffy and gray bands (see Fine Points) and a white terminal band. **Prairie Merlins** are much paler and somewhat larger than other Merlins, and their underwings appear much paler. They have more and larger pale spots on dark primaries and secondaries. **Prairie adult male.** Overall pattern is like Taiga adult male, but much paler. Mustache mark is faint or absent. Streaked crown, back, and upperwing coverts are light blue-gray. Pale areas on hind neck are larger and often have a rufous wash. Whitish underparts are streaked reddish-brown, but with few markings on undertail coverts. Pale rufous leg feathers have fine streaking. Black tail has three or four whitish to light gray bands and a wide white terminal band. **Prairie adult female.** Overall pattern is like Taiga adult female, but much paler. Crown, back, and upperwing and uppertail coverts are medium brown with a grayish cast. Creamy underparts are streaked reddish-brown, but with few markings on undertail coverts. Medium brown tail has three or four wider whitish bands and a white terminal band. **Prairie juvenile.** Similar to Prairie adult female, but back and upperwing and uppertail coverts lack the grayish cast. Tail is like that of Prairie adult female.

UNUSUAL PLUMAGES There is a report of a mostly white bird that also had some tan feathers. There are records of partial albinism in the European race. An adult male and a female captured for banding in Virginia were amelanistic.

FINE POINTS Tail bands of Taiga juveniles are buffy and often incomplete on females, and buffy and grayish and more noticeable on males.

ETYMOLOGY *Falco* is from the Latin *falx*, "sickle," in reference to the falcon's wing shape in flight or, according to another source, to the shape of their beak and talons. *Columbarius* in Latin means "pertaining to a dove (pigeon)," a reference to the Merlin's original North American common name of Pigeon Hawk, after its resemblance in flight to a pigeon. "Merlin" derives from the Old French *esmerillon*, the name for this species.

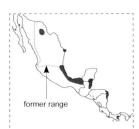
APLOMADO FALCON PLATE 29

Falco femoralis

HALCÓN PLOMIZO

LENGTH 35–45 cm   **WINGSPREAD** 78–102 cm   **WEIGHT** 208–360 g

IDENTIFICATION SUMMARY Colorful, narrow-winged, long-tailed, medium-sized falcons. Long tail, bold face pattern, and dark cummerbund are distinctive. Wingtips extend three-quarters of the way down the tail on perched falcons.

TAXONOMY AND GEOGRAPHIC VARIATION *F. f. septentrionalis* occurs north of Panama, with no geographic variation. Nominate subspecies occurs in Panama, but they hardly differ from falcons farther north. One other subspecies occurs in the Andes of South America.

SIMILAR SPECIES (1) **American Kestrel** (plate 29) is smaller, has rufous tail and back and two dark mustache marks (vertical dark bars) on each cheek, and lacks dark cummerbund. In flight, tail appears shorter and underwings paler than those of Aplomado.

Aplomado Falcon. Adult. Adults have lead-colored upperparts and show white bars on the belly band. Texas USA. December 2010. WSC

284

(2) **Merlin** (plate 29) is smaller, has completely streaked underparts, and lacks distinct face pattern and dark cummerbund. In flight, tail appears shorter than that of Aplomado.

(3) **Bat Falcon** (plate 30) appears similar; see under that species for distinctions.

(4) **Peregrine Falcon** (plate 31) is larger, has broader wings and a single thick mustache mark, and lacks dark cummerbund.

(5) **Prairie Falcon** (plate 31) is larger, paler, and browner; has broader wings with a black center on underwing and a plain pale tail; and lacks dark cummerbund.

(6) **Mississippi Kite** (plate 8) is similar in size and silhouette and also has light line on trailing edge of wing, but lacks strong face pattern and cummerbund.

STATUS AND DISTRIBUTION Uncommon and very local in c Chihuahua, along the Caribbean slope of s Mexico, s Yucatán Peninsula, n Belize, e Guatemala, se Honduras, ne Nicaragua, and on the Pacific slope in c Panama.

Aplomado Falcon. Adult. All show whitish breast, dark belly band, and rufous belly and leg feathers. Adults have white barring in the belly band. Adult females have narrow dark breast streaks, absent on males. Note white band on the trailing edge of secondaries. Texas USA. June 2002. wsc

Aplomado Falcon.
Juvenile. Juveniles lack
white bars on the belly
band and show wide
dark streaks in the
breast. Texas USA. July
2006. wsc

HABITAT Occurs in a variety of open habitats, including open savannah, wet and dry grasslands, and semi-desert.

BEHAVIOR They feed primarily on large flying insects and birds, which they capture after a rapid direct flight from a perch, sometimes including a long tail-chase or pursuit in heavy brush on foot. They hunt from both exposed and inconspicuous perches. They also take reptiles, rodents, and even bats, and they regularly pirate food from other raptors and egrets. Hunting from a soar and cooperative hunting by pairs have also been reported. They regularly gather at grass fires to hunt displaced birds, insects, reptiles, and mammals.

Active flight is rapid and direct, with light, quick wing beats, but when a bird is not pursuing prey, flight is slower, buoyant, and kestrel-like. They soar on level wings and glide on flat wings or with wrists below the body and wingtips curved upward. They will hover for a short time when prey goes undercover.

Breeding pairs remain together throughout the year and often perch close together. They use the stick nests of other species.

MOLT Annual molt is complete. Juveniles begin molt into adult plumage in their first spring.

DESCRIPTION Sexes are almost alike in plumage, but female is noticeably larger. Juvenile plumage is similar to that of adult. Eyes are dark brown. **Adult.** Head has distinctive pattern of lead-gray crown, black line behind eye, and thin black mustache mark; and creamy to whitish superciliary lines, which join together to form a V on the hind neck; and a creamy to whitish cheek and throat. Back and upperwing coverts are lead-gray. Trailing edge of dark wing has noticeable pale band that extends from body to primaries. Whitish to rufous breast is unmarked on older adult males and has a few short, narrow dark streaks on females and first-plumage adult males; blackish cummerbund shows narrow white barring. Underwings appear dark. Leg feathers and undertail coverts are rufous. Long black tail has seven or more thin white bands. Cere, eye-ring, and leg colors are yellow. **Juvenile.** Similar to adult, but back has a brownish cast and buffy feather edges, buffy breast shows more and wider dark streaks, and dark cummerbund lacks white barring. Dark tail has nine or more thinner buffy bands. Cere, eye-ring, and leg colors are pale yellow to yellow.

FINE POINTS Adult's cummerbund has fine white barring, heavier on male, faint or lacking on juvenile.

UNUSUAL PLUMAGES An amelanistic falcon was seen and photographs taken in s South America.

HYBRIDS No hybrids have been reported

ETYMOLOGY *Falco* is from the Latin *falx*, "sickle," in reference to the falcon's wing shape in flight or, according to another source, to the shape of its beak and talons. The species name, *femoralis*, is Latin, "referring to the thighs," for the rufous coloration of the leg feathers. *Aplomado* in Spanish means "lead-colored," a reference to this species' back color.

BAT FALCON PLATE 30
Falco rufigularis

HALCÓN MURCIELAGUERO

LENGTH 24–29 cm   **WINGSPAN** 52–65 cm   **WEIGHT** 118–242 g

IDENTIFICATION SUMMARY Small, dark, compact falcons. Overall, they are dark slate with white throat and sides of neck, white barring on belly, and rufous legs and undertail coverts. They sit on high elevated perches looking for prey, which they pursue quickly and directly. They can appear somewhat swift-like when gliding as their wings are held below the horizontal. Wingtips fall just short of tail tip on perched falcons.

TAXONOMY AND GEOGRAPHIC VARIATION *F. r. petoensis* occurs throughout, with no geographic variation. Differs from nominate race, which occurs in e South America, by slightly paler slate upperparts. Subspecies *petrophilus* of nw Mexico is not valid.

SIMILAR SPECIES (1) **Orange-breasted Falcon** (plate 30) adult is very similar to Bat Falcon adult, but is larger, with proportionally larger head, beak, and feet. It always shows a wide rufous band below the white throat. Juvenile Orange-breasted Falcons differ more in plumage and are less adult-like. Juveniles are overall browner with little rufous on legs and breast.

(2) **White-throated Swift** (*Aeronautes saxatalis*) **and White-collared Swift** (*Streptoprocne zonaris*) in flight at a distance can

Bat Falcon. Adult. Adults have a narrow rufous band between the white throat and dark breast. Note the large head and long narrow toes. Belize. February 2007.

RYAN PHILLIPS—BRRI

appear much like Bat Falcon, but have dark throat and proportionally longer wings.

(3) **Merlin** (plate 29) in flight is somewhat similar; see under that species for distinctions.

STATUS AND DISTRIBUTION Fairly common to common breeding residents on both slopes from n Mexico to Panama, but absent from Mexican plateau, Pacific slope of El Salvador and Nicaragua, and mountains of s Mexico to n Costa Rica.

Bat Falcon. Adult. Adults have a narrow rufous band between the white throat and dark breast. All show dark underwings and dark tail with narrow white bands. Costa Rica. August 2011.
CHRIS JIMÉNEZ

Bat Falcon. Juvenile. Juveniles have streaks in the upper breast and a brownish cast to the breast. Undertail coverts (not shown) show dark barring. Costa Rica. April 2004. wsc

HABITAT Occurs at edges of lowland and foothill rain forest, and cleared areas with some trees, including towns.

BEHAVIOR They prey on flying birds, bats, and insects, but occasionally small mammals and reptiles. They search for prey from an elevated, usually exposed perch and pursue prey with rapid direct flight.

Powered flight is rapid and direct. They soar and glide with wings held below level, appearing much like a large swift. They soar only occasionally.

They usually nest in cavities, but also on cliffs, abandoned buildings, and the crown of palms.

MOLT Annual molt is complete. Juveniles begin molt into adult plumage in their first spring. Some juveniles may show significant pre-formative molt.

DESCRIPTION Sexes alike in plumage; female separably larger than male. **Adult.** Head and upperparts are dark slate, with white throat, sides of neck, and upper breast. Blackish lower breast and upperbelly show narrow white barring. Lower belly, leg feathers, and undertail coverts are rufous. Black tail shows three narrow white bands. Black underwings show white spotting, visible only when seen close in good light. Eye-ring and cere are orange-yellow to yellow, and legs are orange to orange-yellow. **Juvenile.** Similar to adults, but with wider rufous-buff feather edges on dark underparts, short dark streaks on breast, wider white tips on tail feathers, and duller rufous, dark-streaked leg feathers and dark-barred undertail coverts. Juveniles have yellow eye-rings, ceres, and legs.

UNUSUAL PLUMAGES No unusual plumages have been described.

HYBRIDS No hybrids have been described.

ETYMOLOGY *Falco* is from the Latin *falx*, "sickle," in reference to the falcon's wing shape in flight or, according to another source, to the shape of the beak and talons. The species name, *rufigularis*, is from the Latin *rufus*, "reddish," and *gularis*, "throat." Common name is after one of its favorite prey.

ORANGE-BREASTED FALCON

PLATE 30

Falco deiroleucus

HALCÓN PECHO ANARANJADO

LENGTH 33–41 cm **WINGSPAN** 69–91 cm **WEIGHT** 330–650 g

former range

IDENTIFICATION SUMMARY Medium-sized, dark, compact rainforest falcon that has a boldly marked and colorful head. They are rare and local. Head and upperparts are dark slate, with white throat extending onto upper breast and sides of neck; orange breast; white barring on dark cummerbund; and orange lower belly, legs, and undertail coverts. They sit on high elevated perches looking for prey, which they pursue quickly and directly, much like a Peregrine Falcon. Wingtips fall just short of tail tip on perched falcons.

TAXONOMY AND GEOGRAPHIC VARIATION Monotypic, with little geographic variation in our area.

SIMILAR SPECIES (1) **Bat Falcon** (plate 30) is very similar to Orange-breasted Falcon adult and somewhat similar to juveniles; see under that species for distinctions.

Orange-breasted Falcon. Adult. Adults have a wide rufous band between the white throat and dark breast. Note the large head and long robust toes. Guatemala. December 2004. WSC

291

Orange-breasted
Falcon. Adult. Adults
have a wide rufous band
between the white throat
and dark breast. All
show dark underwings
and dark tail with narrow
white bands. Belize.
February 2007.
RYAN PHILLIPS—BRRI

STATUS AND DISTRIBUTION Rare and local. Populations in s Mexico and Central America have declined (Berry et al. 2010). They are now recorded only in several areas of Guatemala, Belize, and Panama.

HABITAT Rain forest, especially lowland humid forest, usually near cliffs.

BEHAVIOR They prey on flying birds, especially pigeons and parrots, but also bats and insects, especially at dawn and dusk. They search for prey from an elevated, usually exposed perch and pursue prey with rapid direct flight.

Powered flight is rapid and direct, very Peregrine-like. They soar and glide with wings held level. They soar occasionally.

They usually nest on cliffs, abandoned buildings, and large ruins.

MOLT Annual molt is complete. Juveniles begin molt into adult plumage in their first spring. Some juveniles may show significant pre-formative molt.

DESCRIPTION Sexes are nearly alike in plumage; female separably larger than male. **Adult.** Slate head appears as a dark hood. Throat and upper breast are white. Sides of neck are white and orangish. Rest of breast, lower belly, leg feathers, and undertail coverts are orangish, the latter showing black and white barring. Females often show faint dark streaks on the rufous breast. Cummerbund of blackish upperbelly and lower breast shows noticeable white barring. Upperparts and upperwings are dark black with coverts edged dark blue-gray; broader edging on greater coverts. Uppertail coverts are dark blue-gray, with three or four narrow white bars. Black underwings show white spotting, visible only when seen close in good light. Black tail shows three narrow white bands, which are often not visible. Eye-ring and cere are orange-yellow to yellow, and legs are orange (more often on males) to orange-yellow. Feet are unusually large. **Juvenile.** Somewhat similar to adults, but with much duller

buffy to buffy-rufous breast and vent, brownish cummerbund with buffy feather edges, and darkly marked buffy-rufous leg feathers. Slate upperparts and upperwing coverts show faint narrow feather edges. Buffy-rufous undertail coverts are barred black. Eye-rings and ceres are bluish to dull yellow, and legs are yellow.

FINE POINTS Adult males often show blue-gray and white barring on flanks, visible only close-up.

UNUSUAL PLUMAGES No unusual plumages have been described.

HYBRIDS No hybrids have been described.

ETYMOLOGY *Falco* is from the Latin *falx*, "sickle," in reference to the falcon's wing shape in flight or, according to another source, to the shape of the beak and talons. The species name, *deiroleucus*, is from the Greek *deire*, "throat," and *leukos*, "white."

REFERENCES Berry et al. 2010.

Orange-breasted Falcon. Juveniles. Juveniles are somewhat similar to adults, but have throat and upper breast buffy with narrow dark streaks, dark brown head and upperparts, and blue ceres. Belize. July 2008. YERAY SEMINARIO

PEREGRINE FALCON PLATE 31
Falco peregrinus

HÀLCON PEREGRINO

LENGTH 37–46 cm **WINGSPREAD** 94–116 cm **WEIGHT** 453–1,100 g

(Measurements for *F. p. tundrius; F. p. anatum* is slightly larger)

IDENTIFICATION SUMMARY Large, long-winged dark falcons. Dark hooded head, wide dark mustache stripes, and dark underwings are distinctive. Wingtips extend to, or almost to, tail tip on perched Peregrines.

TAXONOMY AND GEOGRAPHIC VARIATION Two subspecies occur: Tundra Peregrine, *F. p. tundrius*, is paler, and Anatum Peregrine, *F. p. anatum*, is darker.

SIMILAR SPECIES (1) **Prairie Falcon** (plate 31) is paler and has narrower mustache marks than Peregrine, but Tundra juvenile can be similarly pale. Peregrines lack white area between eye and dark cheek found on all Prairies. In flight, Prairies show dark axillaries and median coverts on pale underwings and uppertail paler than back and upperwings. When perched, Prairie's wingtips fall short of tail tip, but Peregrine's reach or almost reach tail tip.

(2) **Mississippi Kite** (plate 8) is similar in shape to flying Peregrines; see under that species for distinctions.

STATUS AND DISTRIBUTION Rare breeding residents in n Mexico on Baja California and on slopes of central plateau and fairly common on migration along the Caribbean shore; less common inland. They winter locally in many areas, particularly near cities.

Peregrine Falcon. Adult *tundrius*. Adults of this race have narrower mustache marks, larger white cheek areas, and white breasts. Wingtips reach tail tip. Mexico. October 2006. wsc

HABITAT Occurs in a wide variety of open habitats, especially in cities.

BEHAVIOR These raptors are awe-inspiring to watch because of their power, grace, and speed in flight. They perform spectacular vertical dives (stoops) from great heights, with wings held tight against the body, diving at and striking birds at high speeds. They eat birds almost exclusively, capturing them usually in the air, but occasionally on the ground.

Active flight is with shallow but stiff and powerful wing beats, similar to those of cormorants. When actively chasing prey, Peregrines often use deeper wing beats. They soar on level wings with widely fanned tail and the outer tail feathers almost touch the trailing edge of the wing, making the tail appear diamond-shaped. Wingtips appear broad and rounded when soaring and narrower and pointed when gliding. They glide with wings level or with wrists below the body and wingtips up. Wingtips bow upward noticeably when falcon is executing a high-speed turn.

They nest primarily on cliffs, but have used stick nests in trees, on building ledges, and, recently, on level sites under bridges. Fledglings often chase after and catch flying insects, such as dragonflies.

Water is no barrier to Peregrines on migration, and they are frequently observed far at sea capturing birds, eating them on the wing, and perching on ships to eat and rest.

MOLT Annual molt of adults is complete, beginning in spring and completed by autumn by many adults, but it is suspended for migration and completed on winter grounds for migratory adults. Adult females begin molt before their mates. Post-juvenile molt begins in

Peregrine Falcon. Adult. Adults have uniformly dark underwings and yellow ceres. Mexico. October 2009. wsc

the spring of the second year and is usually not quite complete, with Tundra falcons completing this molt on the winter grounds. However, some juveniles have an extensive pre-formative molt and have adult-like body plumage.

DESCRIPTION Females are noticeably larger than males. Sexes are almost alike in adult plumage. Juvenile plumage is different from that of adults. **Note.** The forms described below do not relate strictly to races, but are general types that are recognizably different in the field. Falcons of more than one type can occur within a race's breeding range. **Tundra adult.** Slate-gray head has pale forehead, dark mustache mark narrower than those of other forms, large white area on cheek, and white throat. Slate-gray back and upperwing coverts have blue-gray barring and fringes, heavier on lower back. Uppertail coverts appear a paler blue-gray. Upperwings are slate-gray. Underwing coverts are evenly barred black and white, and slate undersides of flight feathers have some pale barring. White underparts have, at most, a faint rufous wash on the belly, heavier on females. Breast is unstreaked or lightly streaked, heavier on females; flanks and undertail coverts have black barring, heavier on females; and belly shows black spots. White leg feathers are finely barred black. Blackish uppertail has eight or more narrow blue-gray bands and a wide white terminal band, and contrasts with paler blue-gray uppertail coverts. Adult eye-ring, cere, and leg colors are yellow to yellow-orange, brighter on males. ***Anatum* adult.** Similar in plumage to Tundra adult, but head is blackish with dark forehead, wider mustache mark, much smaller whitish area on cheeks (sometimes absent), and usually a stronger rufous wash on underparts. Back and upperwing coverts are more blackish. **Tundra juvenile.** Head has distinctive pattern: buffy forehead, dark brown crown, buffy superciliary lines, dark eye-lines, and buffy cheek and throat separated by narrower dark mustache marks. Some juveniles have completely or mostly buffy crowns and narrower dark eye-lines and mustache marks. Dark brown back and upperwing and uppertail coverts have wide pale buffy to buffy-rufous feather edges; some paler juveniles have scaly upperparts. Upperwings appear dark brown. Dark brown underwing coverts have buffy markings, and dark brown undersides of flight feathers have buffy banding. Creamy underparts are narrowly streaked dark brown. Creamy leg feathers have narrow dark streaks. Creamy undertail coverts are barred or streaked dark brown. Dark brown uppertail has 10 or more usually incomplete narrow buffy to rufous bands and a wide whitish terminal band, with bands conspicuous on paler falcons. Juvenile's tail is longer and wings are wider than those of adults. Juvenile eye-rings and ceres are pale blue-gray, occasionally dull yellow; leg color varies from pale blue-gray to yellow. ***Anatum* juvenile.** Similar to Tundra juvenile, but head is mostly dark with smaller rufous-buff cheek patches, upperparts have fewer and narrower tawny edges, streaking on rufous-buff underparts is wider and heavier, and dark markings on leg feathers are chevron-shaped. Undertail coverts are usually darkly barred.

FINE POINTS Peregrines have proportionally longer primaries than do other North American falcons: as a result, the bend at the wrist appears in flight relatively closer to the body.

UNUSUAL PLUMAGES A Tundra juvenile captured for banding in Texas was overall café au lait in color (amelanistic), with narrow mustache marks noticeable; two other specimens of amelanistic juveniles have many cream-colored (normally dark brown) feathers. Sight records exist of birds with a few white feathers. Partial albinism has been reported from the British Isles.

HYBRIDS Natural hybrids with Prairie Falcons have been reported, but no descriptions were given of the offspring. Hybrids captive-bred with several species of large falcons for falconry sometimes escape.

ETYMOLOGY *Falco* is from the Latin *falx*, meaning "sickle," in reference to the falcon's wing shape in flight or, according to another source, to the shape of the beak and talons. The species name, *peregrinus*, is from the Latin for "wandering," for their long-distance migrations and dispersals. Formerly called Duck Hawk in North America, presumably for one of its prey.

Peregrine Falcon. Juvenile *anatum*. Juveniles of this race are darker and more heavily marked. Juveniles' ceres are blue becoming yellowish with age. Mexico. March 2014. ANTONIO HIDALGO

PRAIRIE FALCON PLATE 31
Falco mexicanus
HALCÓN PRADEÑO

LENGTH 37–47 cm   **WINGSPREAD** 90–114 cm   **WEIGHT** 420–1,140 g

IDENTIFICATION SUMMARY Large, pale, long-tailed falcons. White area between eye and dark ear patch and dark axillaries are distinctive. Large head appears blockish. Wingtips fall short of tail tip on perched falcons.

TAXONOMY AND GEOGRAPHIC VARIATION Monotypic, with no geographic variation.

SIMILAR SPECIES (1) **Peregrine Falcon** (plate 31), especially juvenile Tundra Peregrines, is similar in size and appearance; see under that species for distinctions.

(2) **Swainson's Hawk** (plate 19), especially pale juveniles, have a similar face pattern and appear similar to Prairie Falcons when perched; see under that species for distinctions.

(3) **Prairie Merlin** (plate 29) is much smaller, has faint mustache marks, has noticeable pale bands in tail, and lacks dark axillaries.

STATUS AND DISTRIBUTION Uncommon to fairly common breeders on cliffs in n Mexico (n Baja California and central plateau), moving into level lands when not breeding.

HABITAT Occurs in semi-arid to arid areas, including grasslands and rocky deserts.

Prairie Falcon. Adult. Adults have yellow ceres and white underparts marked with short narrow dark streaks on the breast, dark spots on the belly, and dark barring on the flanks. Mexico. January 2014. ANTONIO HIDALGO

BEHAVIOR They hunt from either a high perch or in level flight. Their prey are small mammals, especially ground squirrels, and ground-dwelling birds, especially Horned Larks (*Eremophila alpestris*), but they also take lizards and flying insects. They fly fast and low over open country and surprise prey, which they capture as the prey attempts to escape. From either perch or soar, they stoop to pick up speed and then close rapidly on prey in a ground-hugging flight. Birds that flush are often tail-chased a considerable distance. Unlike Peregrines, they readily take prey on the ground. On occasion, they hover, often for many seconds, looking for prey that was lost from sight.

Active flight is with shallow, stiff, powerful wing beats, with wings mostly below the horizontal. They soar on level wings, with the tail somewhat fanned, and glide on level wings or with wrists below the body and wingtips curved upward.

They nest almost exclusively on cliffs, but tree nests and a building ledge nest have been reported. During winter they inhabit areas where birds concentrate.

MOLT Annual molt of adults is complete, beginning in May and completed by October. Post-juvenile molt begins earlier in the spring and is also completed by autumn.

DESCRIPTION Adult and juvenile plumages are similar. Sexes are almost alike in plumage; females are noticeably larger. Eyes are dark brown. **Adult.** Brown head has pale superciliary lines, narrow dark eye-lines, pale markings on hind neck, and whitish cheeks and throat separated by narrow dark mustache marks. Small white area between eye and dark ear patch is unique. Back and upperwing coverts are medium brown, with buffy bars and edges on most feathers, and appear overall sandy, paler than those of juveniles, with males averaging paler than females. Underwings appear pale except for dark brown axillaries and median coverts; the latter are paler on males. Whitish underparts are marked with a few short streaks on the breast, rows of spots on the belly, and barring on flanks. Whitish leg feathers have dark spots or bars. Long tail often appears paler on uppersides compared to back and upperwings, and shows incomplete buffy banding. Cere, eye-rings, and legs are yellow-orange, averaging brighter on males. **Juvenile.** Similar to adult, but back and upperwing coverts appear darker because most feathers lack buffy barring. Buffy underparts are heavily marked with dark brown streaks and fade to creamy by spring, but undertail coverts are unmarked. Creamy leg feathers have narrow dark streaks. Uppertail is brown and unbanded, with wide white tip, but is paler below with narrow dark bands. Cere and eye-rings are blue-gray, becoming pale yellow by spring. Legs are pale lead-gray, becoming pale yellow by spring.

FINE POINTS Prairie Falcon's eyes are larger and the tail is longer than that of Peregrines. First-plumage adults, especially males, have

Prairie Falcon. Adult female. Adult females have dark median coverts that show as a band across the underwings. Those of males are paler. Texas USA. January 2009.
GREG LASLEY

a grayish cast to upperparts, lacking on juveniles and usually lacking on older adults.

UNUSUAL PLUMAGES Records exist of two partial albinos and an amelanistic adult female with some cream-colored and some normal feathers.

HYBRIDS Natural hybrids with Peregrines have been reported, but no descriptions were given of the offspring. Captive-bred adult and juvenile hybrids of Peregrine x Prairie Falcons often appear most like a Peregrine, especially the juvenile plumage.

ETYMOLOGY *Falco* is from the Latin *falx*, meaning "sickle," in reference to the falcon's wing shape in flight or, according to another source, to the shape of the beak and talons. The species name, *mexicanus*, is for Mexico, where the first specimen was collected.

REFERENCES

Amadon, D. 1982. "A Revision of the Sub-buteonine Hawks (Accipitridrae, Aves)." *American Museum Novitates* 2741:1–20.

Amaral, F. R., F. H. Sheldon, A. Gamauf, E. Haring, M. Riesing, L. F. Silveira, and A. Wajntal. 2009. "Patterns and Processes of Diversification in a Widespread and Ecologoically Diverse Avian Group, the Buteonine Hawks (Aves, Accipitridae)." *Molecular Phylogenetics and Evolution* December. Doi:10.1016/j.ympev.2009.07.020.

Berry, R. B., C. W. Benkman, A. Muela, Y. Seminario, and M. Curti. 2010. "Isolation and Decline of a Population of the Orange-breasted Falcon." *Condor* 112:479–489.

Bloom, P., and W. S. Clark. 2001. "Molt and Sequence of Plumages of Golden Eagles, and a Technique for In-hand Ageing." *North American Bird Bander* 26(3):97–116.

Clark, W. S. 1999. *A Field Guide to the Raptors of Europe, the Middle East, and North Africa.* Oxford, UK: Oxford University Press.

———. 2004a. "Wave Molt of the Primaries of Accipitrid Raptors, and Its Use in Ageing Immatures." *Raptors Worldwide*, R. D. Chancellor and B.-U. Meyburgh, eds., Proceedings of the V World Conference on Birds of Prey, held in Budapest, Hungary, May 2003, World Working Group on Birds of Prey, Berlin, Germany.

———. 2004b. "Is the Zone-tailed Hawk a Mimic?" *Birding* 36:494–498.

———. 2007. "Taxonomic Status of the Mangrove Black-hawk *Buteogallus (anthracinus) subtilis.*" *Bulletin of the British Ornithologists' Club* 127(2):110–117.

———. 2008. "Hybrid Bald Eagle × Steller's Sea Eagle from Vancouver Island, British Columbia." *Birding* 40:28–31.

Clark, W. S., and R. Banks. 1992. "The Taxonomic Status of the White-tailed Kite." *Wilson Bulletin* 104:571–579.

Clark, W. S., and B. K. Wheeler. 2001. *A Field Guide to Hawks (of) North America.* Revised. Peterson Field Guide series, no. 35. Boston: Houghton Mifflin.

Clark, W. S., and P. H. Bloom. 2005. "Plumages of Basic II and Basic III Rough-legged Hawks." *Journal of Field Ornithology* 76:83–89.

Clark, W. S., M. Reid, and B. K. Wheeler. 2005. "Four Cases of Hybridization in North American Buteos." *Birding* 37:256–263.

Clark, W. S., H. L. Jones, C. Benesh, and N. J. Schmitt. 2006. "Field Identification of the Solitary Eagle." *Birding* 38(6):66–74.

Clark, W. S., and C. C. Witt. 2006. "First Known Specimen of a Hybrid *Buteo*: Swainson's Hawk (*Buteo swainsoni*) × Rough-legged Hawk (*B. lagopus*) from Louisiana." *Wilson Journal of Ornithology* 118:42–52.

Clark, W. S., and P. Pyle. 2015. "Commentary: A Recommendation for Standardized Age-class Plumage Terminology for Raptors." *Journal of Raptor Research* 49(4):513–517.

Edelstam, C. 1984. "Patterns of Molt in Large Birds of Prey." *Annales Zoologici Fennici* 21:271–276.

Ferguson-Lees, J., and D. A. Christie. 2001. *Raptors of the World.* London: Christopher Helm.

Gunther, K., and M. E. Hopey. 2010. "Apparent Northern Goshawk (*Accipiter gentilis*) × Cooper's Hawk (*A. cooperii*) Hybrid in Eastern Tennessee." *North American Birds* 64:524–526.

Hull, J., W. Savage, J. P. Smith, N. Murphy, L. Cullen, A. C. Hutchins, and H. B. Ernest. 2007. "Hybridization among Buteos: Swainson's Hawks (*Buteo swainsoni*) × Red-tailed Hawks (*Buteo jamaicensis*)." *Wilson Journal of Ornithology* 119(4):579–584.

Jenner, T. 2010. "Life History of the White-breasted Hawk (*Accipiter chionogaster*)." *Ornitología Neotropical* 21:157–180.

Jollie, M. 1947. "Plumage Changes in the Golden Eagle." *Auk* 64:549–578.

Liguori, J., and B. L. Sullivan. 2010. "Comparison of Harlan's Hawks with Western and Eastern Red-tailed Hawks." *Birding* 42:30–37.

Miller, A. H. 1941. "The Significance of Molt Centers Among the Secondary Remiges in the Falconiformes." *Condor* 43:113–115.

Millsap, B., S. Seipke, and W. S. Clark. 2011. "Gray Hawk (*Buteo nitidus*) Is Two Species." *Condor* 113(2):326–339.

Morrow, J., L. Morrow, and T. G. Driscoll. 2015. "Aberrant Plumages in Cooper's Hawks." *Journal of Raptor Research* 49:501–505.

Oatley, G., R. E. Simmons, and J. Fuchs. 2015. "A Molecular Phylogeny of the Harriers (*Circus*, Accipitridae) Indicates the Role of Long-distance Dispersal and Migration in Diversification." *Molecular Phylogenetics and Evolution* 85:150–160.

Olson, S. L. 2006. "Reflections on the Systematics of *Accipiter* and the Genus for *Falco superciliosus* Linnaeus." *Bulletin of the British Ornithologists' Club* 126:69–70.

Phillips, R., and Y. Seminario. 2009. "Aging Characteristics of *Spizaetus melanoleucus*: First Photographic Documentation." *Neotropical Raptor Network Newsletter* 9.

Pyle, P. 2005. "First Cycle Molt in North American Raptors." *Journal of Raptor Research* 39 (4):378–385.

Schroeder, R. 2009. "Melanistic Cooper's Hawk in Lyon County." *Loon* 81:96–97.

Seipke, S. H., A. M. Castaño, and K. L. Bildstein. 2007. "Spanish Common Names of Raptors in Latin America." In *Neotropical Raptors*, K. L. Bildstein, D. R. Barber, and A. Zimmerman, eds., Orwigsburg, PA: Hawk Mountain Sanctuary.

INDEX